Histology, Immunohistochemistry and In Situ Hybridisation.
Lab protocols.

Jenny Steel

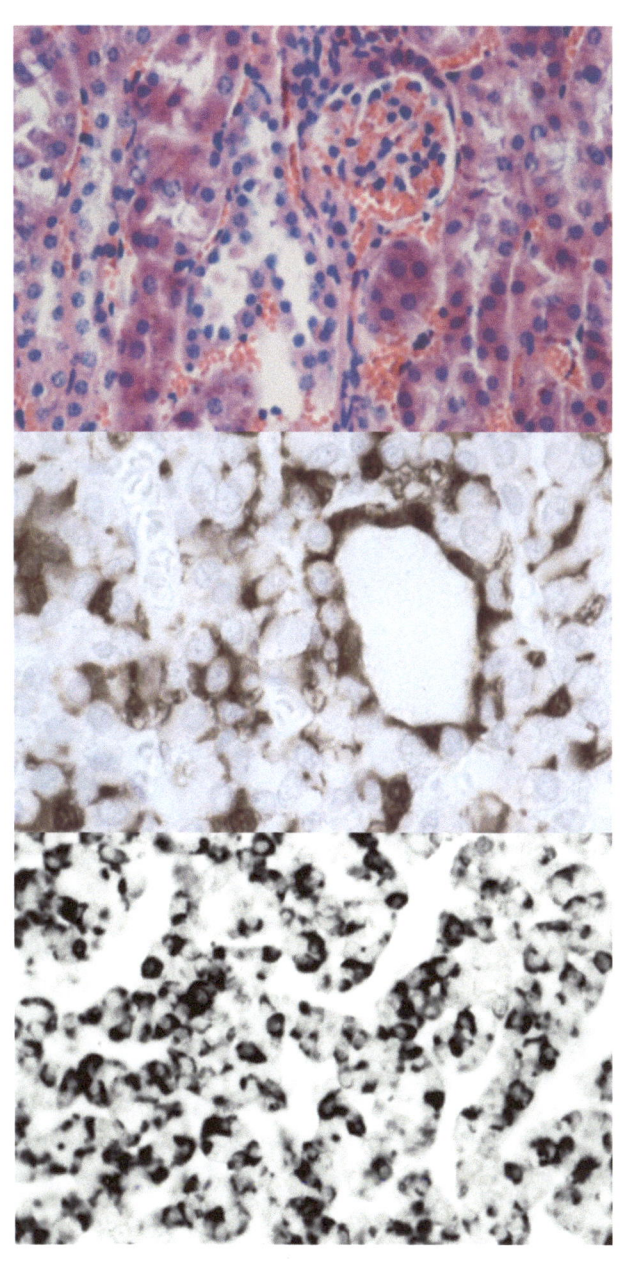

© 2017 Jenny Steel. All rights reserved.
ISBN 978-1-326-98815-9

Preface.

This book is based upon the methods which I have used successfully over my career. I learned about histology at university and then in a routine diagnostic cytology/histology lab. I used histological methods, immunohistochemistry (IHC) and in situ hybridisation (ISH) during my PhD in the Department of Histochemistry at Hammersmith Hospital, with Professor Julia Polak. This department was founded by Professor A.G.E. Pearse, and his department was renowned for the development of IHC methods. In Julia's lab, we were pioneers of using ISH for detection of mRNA in cells and tissues, and I published some papers showing what the technique could do. Most of our early ISH work was done using radiolabelled probes and frozen sections, but later (mainly at the ICRF) we demonstrated that paraffin sections were just as good as frozens, more versatile, and that digoxigenin-labelled probes were safer and easier to use than radioactive probes.

The aim of this book is to share my own experience of techniques which have worked in my hands, rather than give an overview of available methods. Techniques used in histology, IHC and ISH are available widely in books and websites - there is so much information out there that it can be confusing for a beginner. I hope that this slim volume might help with some pointers and a place to start from.

There are references at the end - the intention is to provide evidence that the techniques work, rather than write a review, so these are references to my own work.

Personal note:
I was diagnosed with cancer in 2016 and although I had spent most of my working life involved in research into tumours of various types, this didn't prevent me from becoming a cancer patient – no one is immune from this disease, unfortunately. I took early retirement in January 2017, on the grounds of ill health. This book is an attempt to contribute something extra at this stage of my life, after the phase when I was able to conduct research projects myself.

JS 2017

Contents: Page

Preface — 3

Paraffin embedding for formalin-fixed tissues. — 5

Cutting paraffin sections. — 8

Haematoxylin and eosin (H&E) staining. — 11

Optimisation of antibodies for immunohistochemistry — 13

Immunohistochemistry with ABC peroxidase. — 14

Examples of immunohistochemical results — 17

Controls for immunohistochemistry — 19

Troubleshooting immunohistochemistry. — 20

Supra-optimal dilution technique for immunohistochemistry — 21

Co-localisation methods using immunohistochemistry — 23

In situ hybridisation - introduction. — 25

Diagrammatic description of ISH procedure — 27

Why use digoxigenin? — 28

Equipment and reagents for ISH. — 29

Controls for ISH. — 31

Detailed protocols for ISH. — 32

Examples of ISH results. — 43

Troubleshooting for ISH. — 44

Simultaneous IHC and ISH — 45

In situ RT-PCR — 47

References — 50

Paraffin embedding for formalin-fixed tissues.

All procedures carried out at room temperature unless indicated. Tissue should be immersed in about 10 times its volume of liquid at each step. Plastic universals can be used for all the steps apart from Histoclear.

Fixation.

Fix tissue in Neutral Buffered Formalin or 4% paraformaldehyde for up to 24 hours. Most tissues can be fixed overnight. The purpose of fixation is to preserve the structure of the tissue, and unless this is carried out adequately, it will not be possible to see the cells clearly or detect proteins immunohistochemically.
For small tissues such as mouse ovary or small biopsies, fix them for about 6-8 hours. If tissues are fixed for too long they can become hard and it is difficult to reverse this process.
Formalin fixed, paraffin embedded (FFPE) tissues are versatile and can be used for many different techniques, including in situ hybridisation, microdissection methods and DNA extraction.

Processing tissues.

Following fixation, rinse fixed tissues in 70% ethanol and store them in 70% ethanol (no time limit). Formalin must be discarded into chemical waste.

For processing to embed in paraffin:
Place tissue samples in glass bottles containing fresh 70% ethanol and insert a small paper label along with the tissues, with the number or identity of the specimen written in pencil (not ink). Several pieces from the same specimen can be processed in the same bottle. If all the tissues are the same and do not need to be identified, there is no need for a label. The labels should remain with the specimen at all stages of the procedure, and will prevent confusion of specimens once they are removed from the bottles later.

Dehydrate tissue by removing the 70% ethanol and adding absolute ethanol etc, according to the following approximate timetable (pour liquid out into a beaker or use a disposable Pasteur pipette to remove the solvents). Timing is not critical – you can leave tissue in ethanol for longer, or for small pieces of tissue reduce time to 40 minutes per step.

Absolute ethanol - 3 changes of 1 hour each.
Histoclear (in a glass container) - 2 or 3 changes of 1 hour each.

Tissues can be left overnight in Histoclear. The tissue will "clear" or become translucent in the first change of Histoclear. If clearing does not occur, the tissue must return to absolute ethanol again until thoroughly dehydrated.

Embedding.

Once tissues have been cleared in 2 changes of Histoclear, they can be embedded in wax.
Wax is melted in a wax dispenser at 60-65°C. You will also need a dedicated wax oven set at about 60°C (up to 65°C).

Cover the bench with paper towels to collect drips – wax is difficult to remove from the bench.
Protect yourself against hot wax – wear gloves if necessary.
A standard oven will drop in temperature rapidly if the door is left open.

Heat up one or more clean 6-, 12- or 24-well plastic culture plates to 60°C for about 10 minutes before use and then fill the wells with wax using the tap on the wax dispenser, depending on the number of samples to embed and the size of the samples, allowing enough wax to fully immerse them in the well.

Working quickly to avoid the wax setting, place one tissue sample with its unique paper label into each well using forceps. Going from Histoclear into wax, ensure specimen is fully surrounded by wax and make sure it doesn't stick to the bottom of the plate (solvent may slightly dissolve the plastic). Drain the excess Histoclear from each piece of tissue onto a paper towel quickly before placing in the wax.

If wax begins to set, place the plate back in the oven for about 10 minutes, then complete the process.

Leave tissue in wax in the oven at 60°C for 1 hour. Initially, the wax will need to melt completely and the samples will need to reach 60°C for the wax to penetrate into the tissue and displace the Histoclear.
At the start of this time, turn on a forceps heating block to heat forceps.
After one hour, transfer tissue samples into fresh wells containing molten wax using hot forceps.
Alternatively, heat up a disposable Pasteur pipette to 60°C and use that to remove the first wax then refill the well with new wax. Leave in fresh wax for 1 hour.
It is necessary to change the wax to remove all of the Histoclear from the wells.

Use clean disposable plastic moulds for embedding.
This step must be done as quickly as possible to prevent wax from setting too soon.
Some people use a hotplate, but it is an advantage if the layer of wax on the base of the mould begins to set, so if you can work quickly this is easier.

Add a small volume of wax from the wax dispenser to each mould in turn and, using heated forceps, quickly transfer a tissue sample from each well into a mould, together with its paper label. Orientate the specimen as you wish.

Place wax-filled mould on a paper towel on the bench, and immediately cover the molten wax with a white plastic cassette lid and press down gently to make a block. Take care that the wax does not spurt through the gaps on the mould – it could end up in your eye! If necessary, wear safety glasses. The wax should come through the holes. If there is insufficient depth of wax in the mould, add some more to the completed block before it sets.

Allow the blocks to set at room temperature. This will take up to an hour, but setting can be speeded up by placing the blocks on ice. If left on ice for too long, cracks can appear in some blocks.

If the tissue is found to be not orientated correctly, the blocks can be placed in the oven to re-melt and then allowed to set again. If in doubt, embed all the tissues quickly and then cut sections from one sample to check its orientation.

When blocks are completely set, remove them from the moulds by pressing gently on the reverse (clear plastic) side of the block. The block should come out cleanly and not leave any wax behind. The blocks are now ready for cutting, or can be stored at room temperature indefinitely.

Write the specimen details on the white plastic lid using lab marker (especially if the paper labels are loose).

Allow the used wax in culture plates to solidify at room temperature and dispose of the plates in the lab bins. Also discard used plastic moulds, paper towels, etc.

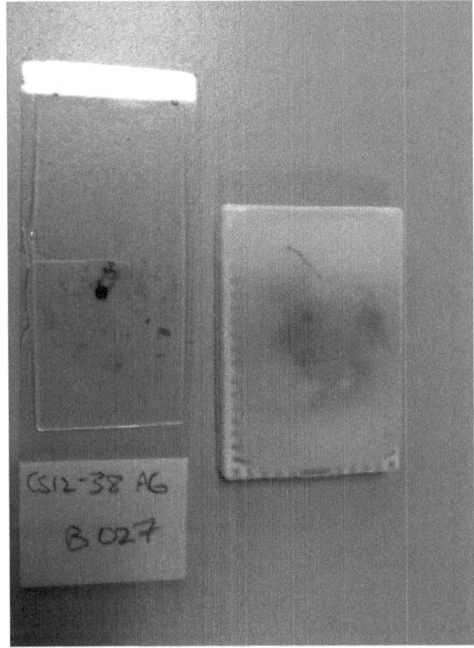

Wax block and an H&E section cut from the block.

Cutting Paraffin sections

This protocol is designed for use with a Leica rotary microtome.

Safety information:
A microtome knife is very sharp indeed and will cut through bone. Always use the brake on the hand wheel when placing a new block into the holder, moving the knife holder, or changing the blade. When cleaning the knife, use a small brush and brush upwards only, to protect the sharp blade from becoming blunt. Avoid wiping the blade with paper tissues.

Before using the microtome, prepare a section-cutting water bath beside it. Make sure that the water bath is full of clean distilled water and turn it on (temperature set to 40°C). If you will use the sections for in situ hybridisation, you may want to fill the water bath with fresh autoclaved distilled water.

Cool your wax blocks before cutting by placing them face upwards on ice or a cold plate for a few minutes, or cooling in a fridge at 4°C. If placed face down on ice, the water could penetrate the tissue and cause artefacts.

Make sure that the chuck with the block holder is wound back using the small winding handle to the left of the base of the microtome. If the block is too far forward, it will not be possible to cut sections.

Make sure the hand wheel brake is on and it cannot move. If there is a knife guard, place it over the knife.
Place a block into the block holder clamp above the knife by pulling the lever at the top towards you. Blocks can be orientated vertically or horizontally in the clamp.

Orientate the block surface so that it is flat against the knife. If the block is at an angle to the blade (vertically or horizontally), you can adjust it by releasing the lever on the left of the block holder itself. Then use the vertical and horizontal adjustment screws to move the block. If using a block cut previously, it is important to cut in the same plane as before, or a large amount of tissue could be wasted.

Move the knife carriage by releasing the lever to the right of the base of the microtome, in front of the hand wheel. Move the knife until it is up against the front of the block. Release the brake if necessary while doing this. Secure the knife holder with the lever so that it cannot move.

Trim the block by turning the hand wheel away from you. You can adjust the section thickness to trim more quickly or use the trim lever on the left of the microtome, beside the block holder. Take care with blocks containing very small specimens because enthusiastic trimming could destroy the block! Trim until the surface of the block is flat or until the whole tissue specimen is exposed. If the block was cut previously, only minor trimming is necessary. If in doubt, begin to collect sections when you are not sure whether the tissue is fully exposed yet.

Adjust section thickness to the desired setting (usually 3-5µm) and try cutting a series of sections. Hopefully the sections will stick together so that you can produce a "ribbon", stretching them out gently by holding gently with a pair of forceps (using your left hand). Sections will be flatter if they are stretched like this.

If the sections have holes or vertical score marks, or scrunch up, it is likely that the blade is blunted. Move the knife by releasing the lever at the left side of the knife holder, slide the knife sideways to a fresh area and try cutting again. If the entire blade is blunt, change the blade.
When cutting with a new blade, it is likely that the sections will curl up. Cut a section and try to uncurl it using a small brush, then cut another section and pull the first section away from the knife using forceps. If it is impossible to avoid section curling, reduce section thickness to 3 µm and try again, then when sections are cutting well, you can increase thickness to 5 µm again.

When you have a ribbon of sections, keep hold of the bottom section with forceps and lift the top section from the knife using a small brush or forceps, avoiding damaging the knife edge. Lift the sections across to the warm water bath and lay them flat.

When the sections look flat, separate them into individuals, pairs or whatever number you wish while floating on the water, using forceps. Some types of tissue may require a few minutes to flatten out completely. If sections have wrinkles it should be possible to remove some of these by pressing gently with forceps. Don't be afraid to reject bad sections at this stage because they will not be much use later on.

Mount sections onto clean adhesive-coated glass slides.
Polysine slides are coated with poly-L-lysine which acts as an adhesive. Alternatively, use Superfrost Plus slides which are charged and enhance tissue adhesion. Collect sections from the water bath by inserting a slide vertically into the water close beside the sections and pulling the slide out again, vertically, dragging the sections onto the slide. If the sections don't stick to the slide, use forceps to guide the top edge onto the glass while you pull the slide out of the water.

Drain slides vertically by leaning them against the water bath on a paper towel. When fairly dry, place them on a hotplate or the side of the water bath.

If you want to use sections for immunohistochemistry or in situ hybridisation, dry them at 37°C in an incubator for several hours or overnight, don't melt them at high temperatures. For in situ hybridisation, use RNase-free conditions and wear gloves, keep sections and slides as clean as possible and store in a dust-free box after drying.

Special notes for cutting Tissue Microarrays.
TMAs require special skill because near-perfect sections are required. When using a new blade, blunt it first by cutting an empty wax block. Use a practice TMA before the real ones, to check that the blade is cutting well.

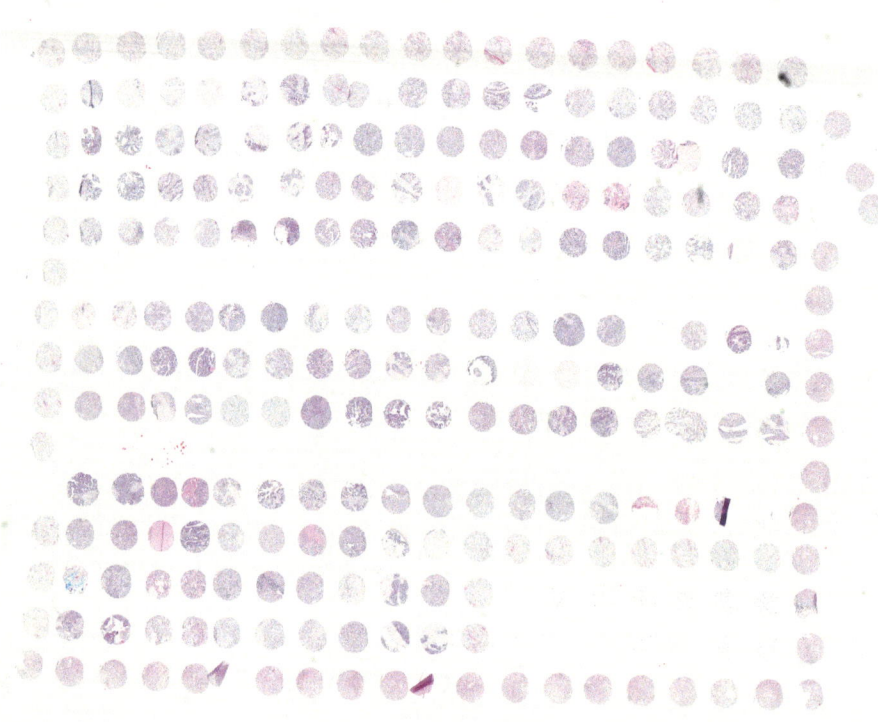

Example of a TMA section.

Haematoxylin and Eosin (H&E) staining for paraffin sections.

Dewax sections:
Dry sections thoroughly in an oven at 37°C.
Place slides into a slide rack.
Immerse slides in Histoclear (2 changes), absolute ethanol (2 changes), 70% ethanol and then distilled water. Each step takes about 1 minute.

Stain in Haematoxylin for about 5 min. (stains nuclei blue)

Rinse in tap water until water is clear.

De-stain in 1% acid alcohol for a few seconds only.

Rinse in tap water until sections appear blue.

Stain in aqueous eosin for about 5 min. (stains cytoplasm red, to various extents)

Rinse in tap water until water is clear.

Drain slides on a folded paper towel on bench to remove excess water from slides before dehydration (don't allow slides to dry).

Rapidly dehydrate – about 10-20 seconds in 70% ethanol, about 30 seconds maximum in the first change of absolute ethanol, 1 minute in second change of absolute ethanol. Leaving slides for too long in ethanols will allow the eosin to be removed from the sections. Some colour will inevitably come out of the sections into the ethanols.

Transfer the slides into Histoclear and move the rack up and down to mix. Leave in Histoclear until the solvent has penetrated the sections and they are translucent. The colours in the sections should now be easy to see. If the Histoclear has droplets in or is cloudy, this suggests that the slides were not rehydrated properly and water is present, which will impair the finished result. In this case, return sections to some clean ethanol, then into clean Histoclear.

Mount slides with coverslips directly from Histoclear in DPX resin mounting medium (care, contains xylene, use fume hood and do not inhale fumes). Do not allow slides to air dry before applying DPX, since this will result in air bubbles being trapped in the tissue. Allow slides to dry in the fume hood face up for at least 30 minutes.

Dry at 37-40°C for about 30 min on paper slide trays, if they need to be viewed quickly. Allow to dry fully at room temperature before storing vertically because the DPX takes time to dry thoroughly.

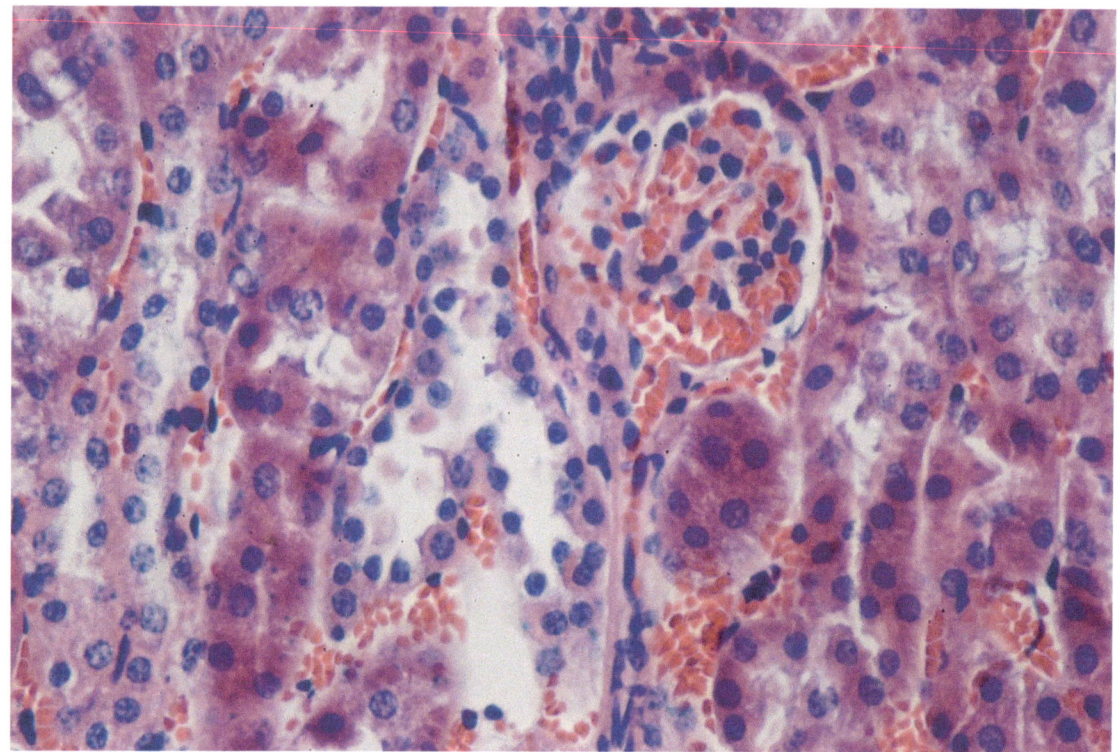
Mouse kidney stained with H&E. Nuclei are stained blue and the cytoplasm red.

Optimisation of antibodies for immunohistochemistry.

Most primary antibodies used for immunohistochemistry (IHC) are commercially available and are supplied with some information about how to use the reagent optimally. Every antibody needs to be optimised for use in the specific tissue or technique in the study in question, because experimental conditions vary and detection limits for antigens may be different from a test section used by the manufacturer.

At optimum dilution, there should be a good signal-noise ratio, with low non-specific background but clear specific staining, which is strong enough to be convincing. This type of staining can often be found at a higher dilution than that recommended. Bearing in mind the cost of primary antibodies, it is more economical to dilute the antibody to its maximum limit. In addition, if any peptide challenge negative controls are required, optimisation is needed to find the maximum dilution for that antibody, so that the staining can be blocked.

For optimisation, it is necessary to begin with basic information such as the species in which the antibody was raised, storage temperature, recommended dilution, antigen retrieval method, etc. If some of these factors are not know, more experimentation will be required.

If the recommended dilution for use of a primary antibody is known, this is a starting point but is not guaranteed to be the optimal choice. With a new antibody it is best to try at least 2-3 serial dilutions, starting from the recommended concentration. For example, if the manufacturer suggests 1:200, try 1:400 and 1:800 as well. The result should be that one dilution is too weak, one is too strong and the middle one is correct. Sometimes, the highest dilution still produces very intense staining, and in this case further dilution should be carried out.

If antigen retrieval is critical to detection of the antigen, this needs to be established before the antibody dilution can be optimised. Again, begin with the recommended pretreatment. Sometimes the amount of time required for pretreatment (eg. microwave treatment with citrate buffer) will need to be altered. If there is no detection of the antigen, another type of antigen retrieval might be required. Buffers at higher or lower pH (pH 6.0, pH 9.0, etc) may be tried, or enzymatic digestion with proteases.

If available, positive control tissues are helpful in checking that primary antibodies are detecting the expected antigen, and they can be used for optimisation as well. If the study involves IHC for proteins expressed at low levels or only in a proportion of the target tissues, positive controls will be needed to ensure that the technique is working correctly and avoid false negatives.

Immunohistochemistry protocol - ABC peroxidase technique.

This method is based upon Vector Laboratories Elite ABC peroxidase kits.

Tissues may be paraffin embedded for wax sections or frozen and cut in a cryostat for frozen sections.
For fixed cells and frozen sections, ie. that do not require removal of wax, omit the first few steps.

Day 1.

1. Paraffin sections - dewax by immersing in Histoclear (2 changes). Remove Histoclear in absolute ethanol (2 changes). Rehydrate in 70% ethanol, then distilled water. One minute in each solvent should be enough.

2. Antigen retrieval:

This is often required for optimum detection of antigens, especially those in cell nuclei. The antigen retrieval recommended for each antibody should be found in notes supplied with the antibody, or can be obtained by asking colleagues. If the appropriate method is unknown, it will be necessary to try various methods initially, using positive control sections.

The usual antigen retrieval method is citrate buffer, pH 6.0.
Make up 0.01M citrate buffer and adjust the pH to 6.0 by adding sodium hydroxide.
(0.1M stock, pH 6.0: 19g citric acid in 1 litre water, pH with NaOH to 6.0.
Dilute x10 with water before use for working solution of 0.01M.)

The dewaxed sections are covered with buffer in a plastic box with a loose-fitting lid, and heated in the microwave for various amounts of time.
For RIP140, usual time chosen is 15 minutes.

Heat for 5 min, check slides to make sure that the liquid is still covering the slides. If the level has gone down, add some distilled water to make up the volume, Heat for another 5 min and repeat, until the total time has elapsed.

Remove the hot slides and buffer with care from the microwave (wear gloves, use protection). Take the slides out of the hot buffer and immerse them in cold distilled water. Some workers advocate allowing slides to cool at room temperature or on ice, but this is a time-consuming method. The main aim of antigen retrieval is to use the same method each time, so cooling slides in cold water is just as good and is quicker.

Alternative antigen retrieval methods:
A) Tris-EDTA pH 9.0 (10 mM Tris base, 1 mM EDTA, in distilled water, pH to 9.0 using NaOH). This is made up fresh and heated to 95°C in a water bath. Slides are dewaxed and added to the solution and left in the water bath for 30 min. The hot glass dish is

then removed from the water bath, and allowed to cool on the bench for 20-30 min. Slides are then rinsed in cold Tris buffer and then in PBS.

B) Enzymatic antigen retrieval. Proteases such as proteinase K, made up to a concentration such as 10-20 µg/ml in buffer (PBS will do). This approach normally requires trying different times for digestion. The solution can be pipetted onto the sections or the slides can be immersed in the solution in a dish.

3. Following antigen retrieval, rinse slides in PBS.

4. Block endogenous peroxidases in the tissue by immersing slides in PBS with 0.3% hydrogen peroxide for 15 min (add 1 ml of 30% hydrogen peroxide to each 100 ml of PBS). For cells grown on coverslips or slides, or frozen sections, use methanol instead of PBS because bubbles in the PBS can lift cells from the glass.

5. Rinse slides 2x 2min in PBS.

6. Incubate sections with normal serum to block non-specific antibody binding. Normal serum from the species in which secondary antibody was raised (e.g. goat, rabbit or horse) is usually supplied in the Vector ABC peroxidase detection kits. Add 75 µl of this serum to 5 ml of PBS (kit specifies 150 µl in 10 ml).

For antibodies which give a lot of non-specific background, use more concentrated blocking solutions, such as 20% normal serum plus 4% BSA.

Before applying the blocking serum, dry each slide carefully around the section using paper tissues, leaving the section moist, or drain excess buffer from slides with cells on. Do not dry the sections themselves.
Draw a circle around the section using a hydrophobic marker pen (eg. Imm-Edge pen, Vector Laboratories), to limit the amount of solution applied to each section.

Place slides in a staining chamber (a plastic box containing paper tissues moistened with water, with a rack to hold slides) so the slides are lying flat. Add a few drops of diluted normal serum to each slide to cover section or cells. Place a lid on the staining chamber to prevent sections from drying out and incubate at room temperature for 20-30 min.

7. Drain serum from sections but do not rinse them. Dry around the sections again.

8. Apply primary antibody, diluted in PBS or in the blocking serum solution used above, to cover sections or cells.

The best dilution to use for the primary antibody needs to be determined for each antibody because specificity will vary. If possible, try several dilutions (each on a separate section) for each new antibody, and try it on a positive control tissue.

Always include a negative control section with no primary antibody, instead use antibody diluent only – normally PBS.
Incubate with primary antibody overnight at 4°C in a moist staining chamber. This step can also be done at room temperature for a shorter time if necessary.

Day 2.

1. Rinse slides twice in PBS (2x 2 min, or longer).

2. While slides are rinsing, dilute secondary and third layer antibodies in PBS:

Secondary antibody - specific for the species in which primary antibody was raised. For example, Vector biotinylated goat anti-rabbit IgG, dilute 1:200 with PBS before use (5 µl per ml of PBS).

Third layer; ABC reagent, made from two reagents, labelled A and B. This will bind to any biotinylated secondary antibody and is not species-specific.
Add 10 µl of each to 1 ml of PBS and leave at room temperature for at least 30 min before use. If made up at the same time as second layer, timing will be ideal.

3. Apply second layer antibody. Pipette onto sections and incubate for 30 min.

4. Rinse slides in PBS (2x 2 min).

5. Apply third layer (ABC) and incubate slides for 30 min.

6. Rinse slides in PBS (2x 2 min).

7. Develop the colour reaction using diaminobenzidine (DAB).

Use a DAB substrate kit (eg. Vector ImmPACT DAB).
To make working solution, add the DAB concentrate, 1 drop per 1 ml of diluent. Wear gloves, DAB is carcinogenic.

Apply DAB solution in drops onto slides in staining chamber and incubate for a short time. The sections will turn brown in colour, possibly very quickly. Start with a 1 minute incubation, rinse slide, examine microscopically, then if necessary add more DAB and incubate for longer.

8. When colour development is complete, rinse slides in running tap water for about 5 min.

9. If desired, counterstain cell nuclei lightly with haematoxylin for up to 30 seconds, rinse in tap water, differentiate in acid alcohol (quick dip), rinse thoroughly in tap water (5 min).

10. Dehydrate sections through ascending alcohols (70% ethanol, 2 changes of absolute ethanol), clear in Histoclear and mount in DPX (contains xylene – use in a fume hood).

Examples of Immunohistochemical staining:

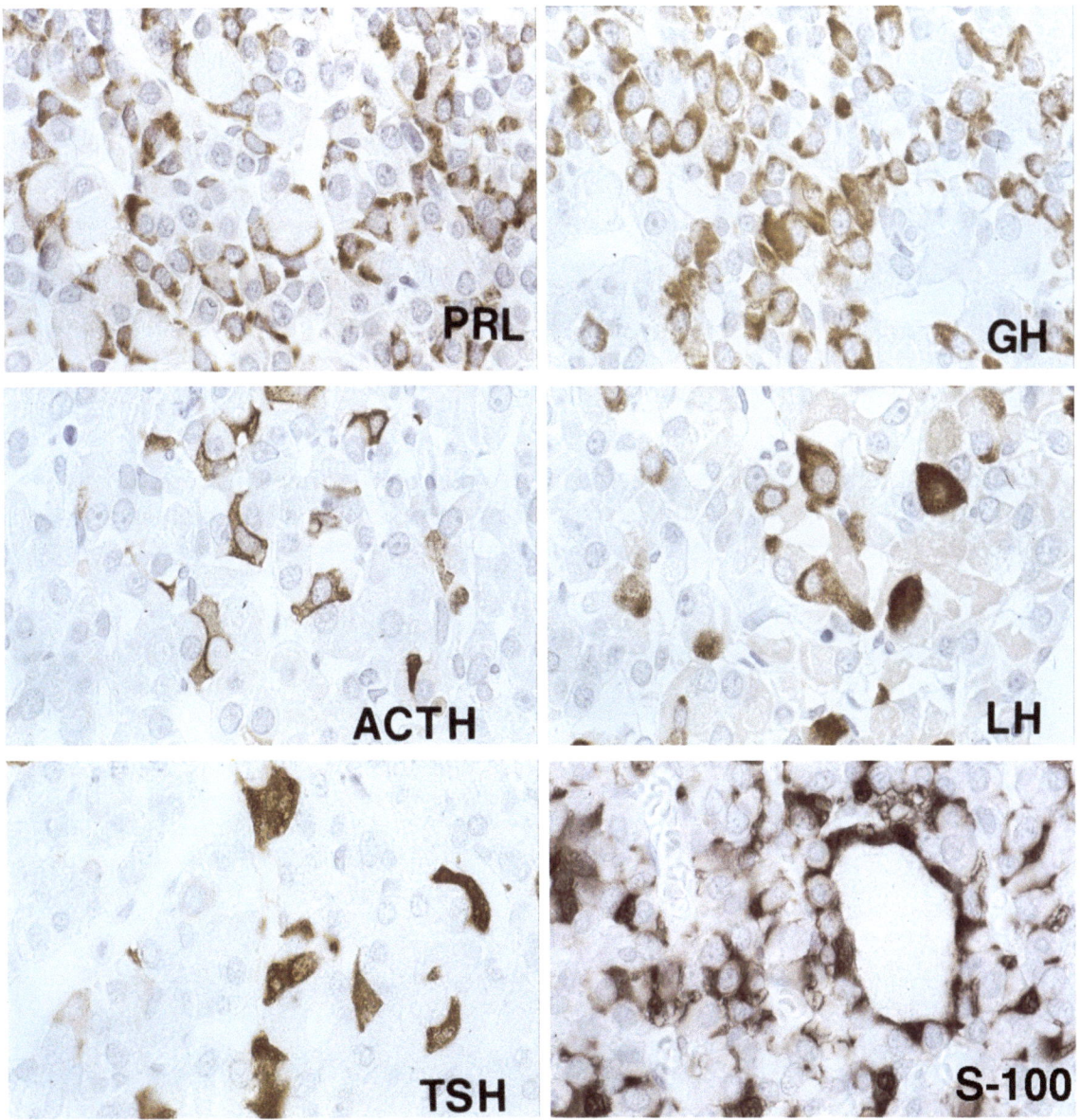

Rat anterior pituitary, showing 6 different types of cells detected using specific antibodies.

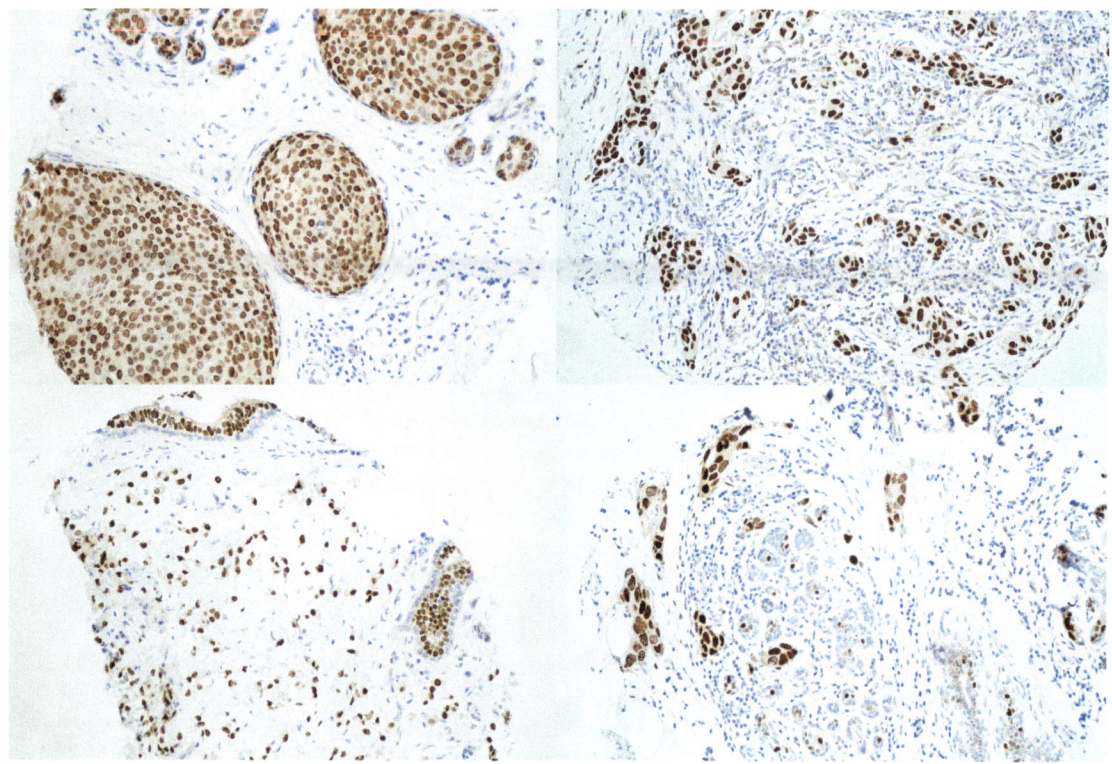

Estrogen receptor alpha (ERα), detected in breast cancer samples in a tissue microarray. Top row shows 2 different tumours with no normal tissue present. Bottom row shows 2 samples containing normal breast epithelium as well as tumour cells.

Controls for immunohistochemistry.

<u>Blocking antibody staining with the specific antigen:</u>
If the primary antibody has been optimised, it will be at a concentration where it can be blocked by the target protein. For a publishable study, it is best to obtain the specific antigen and set up an inhibition experiment.
Use the primary antibody at the highest possible dilution which gives good staining – ie. if 1:1000 stains well, try diluting further to 1:2000 or higher. If you still see good staining, continue to dilute until the staining becomes weaker.

Choose the best optimised dilution and add the antigen to a small quantity of this dilution of antibody. The concentration of antigen to add could be 50-100 nM, this has worked well in the past. Incubate the antigen with the antibody for at least 1 hour before applying to the sections, alongside positive control sections. The idea is that the antigen will bind to the antibody and prevent it from binding to the section.

Make up peptide inhibition tubes as follows (for both specific and unspecific antigens if available):
200 µl of diluted antibody plus 1 µl of a 10 µM dilution of the peptide: 1:200 dilution = 50 nM.
200 µl of diluted antibody plus 2 µl of peptide = 100 nM.
Both concentrations will be added to separate sections, with a third section receiving just antibody.

If inhibition does not work, increase the antigen concentration.

<u>Negative controls:</u>
This is a basic control which should be included in all experiments. Omitting any of the steps in a multi-step method should produce a negative result. Usually, the primary antibody is omitted because if there is any staining in its absence, this suggests the other reagents are not behaving as expected or there may be non-specific staining.

<u>Positive controls:</u>
If in doubt, try to find out what tissues express the target antigen and include some alongside your test samples, if you are testing an antibody on tumours or tissues where it is not known if the antigen is present.
The Human Protein Atlas (www.proteinatlas.org) is a good starting point for human tissues. Any antigen which has been studied in human tissues should be included in the database, and the tissues where it is expressed are listed there.
Normal human tissue samples are often available from histopathology departments, or tissue microarrays of tissues can be purchased. Other types of positive controls include cells transfected with the gene, but these can be unreliable. In some cases, animal tissues can be used as positive controls, although most antibodies are species specific.

Troubleshooting Immunohistochemistry (IHC).

Various things can go wrong with IHC staining. If the method is followed carefully, the main variables which can be changed will be the antigen retrieval conditions and the antibody. The secondary antibody and ABC reagents are stable and usually reliable as long as they haven't expired.

If the entire section is negative the following may be the explanation:
Lack of antigen retrieval if it is required.
Incorrect antigen retrieval conditions.
Mis-match of primary and secondary antibody.
Incorrect primary antibody dilution.
Out of date reagents.

If the entire section is stained uniformly:
Lack of blocking – endogenous peroxidase in particular.
Incorrect blocking serum (eg. blocked with rabbit serum for an anti-rabbit secondary antibody).
Primary antibody concentration too high – dilute it.
Antigen retrieval overdone – try less.
Non-specific staining using the standard protocol – try increasing the blocking method by adding BSA and increasing serum concentration for the primary antibody.
The final explanation is that the antigen is expressed uniformly in all cells.

If there are random areas of staining not related to cell patterns:
Sections may have dried out at some stage.
Over-heating during antigen retrieval and drying out.

Poor quality of tissue structure:
Probably due to poor fixation. Difficult to solve – obtain new samples.

Counterstaining is too pale – repeat it.
Counterstaining is too strong – use acid alcohol to reduce haematoxylin intensity.

Large numbers of samples need to be stained at the same time:
Too many will cause problems. Divide the samples into manageable amounts and make sure that you follow exactly the same methods for each batch.
The DAB substrate development step is critical and so in staining large numbers of slides, set a clock and pipette the DAB reagent onto each slide at 5 or 10 second intervals. Then tip the DAB off each slide after 1 minute, again at 5 or 10 seconds after the previous slide. If in doubt, work out how long it will take to process each slide and set the timings accordingly. If attempting to process a large batch at once, the slides at the end will have more exposure to the DAB than those at the start and won't be comparable in intensity.

Supra-optimal dilution technique for immunohistochemistry.

The dilutions of antisera used in immunocytochemical methods are carefully optimised to reduce nonspecific background while achieving maximum specific staining. The effect of this optimisation is that very small amounts of antigen can be detected, and may give a staining reaction which is almost as intense as that given by larger amounts of antigen. Thus it is difficult to distinguish small differences in antigen concentration. A technique known as the supra-optimal dilution technique makes use of dilutions higher than optimal in order to increase the range of staining intensities obtained (Vacca-Galloway LL, Histochemistry 83:561-9, 1985; Springall DR et al, J Pathol 155:259-67, 1988). Cells containing small amounts of antigen may be negative, whereas those with larger amounts will remain immunoreactive. Thus it is possible to identify samples in which the majority of cells contain less antigen, giving weaker staining, although at optimal dilution these specimens might have stained as intensely as those with more antigen. This method is semi-quantitative and has been used to identify differences in staining which were not detected at the optimal dilution.

This can occur in studies using tissue microarrays (TMAs), comparing biomarker expression which may relate to clinical parameters.

For example, a TMA was stained using an antibody to MSX1 at 1:1500, which had been chosen as the optimal dilution, giving good detection of the antigen. The TMA was scored microscopically by 2 observers and most of the cores were scored as "moderately stained", so this gave no useful information. The antibody was then diluted to 1:3000 and the samples were re-stained. This time, it was clear that some of the samples were very weakly stained, while others remained strongly stained and the "moderate" group became much clearer. The outcome was a clear association between MSX1 expression and survival in this group of patients. Examples are shown on the next page.

If the significance of the expression is unknown, it is difficult to decide what should be the optimum dilution. If it is vital that there should be clear differences, the antibody may need to be diluted to a point where many samples are negative, so that those which express higher levels are easy to spot.

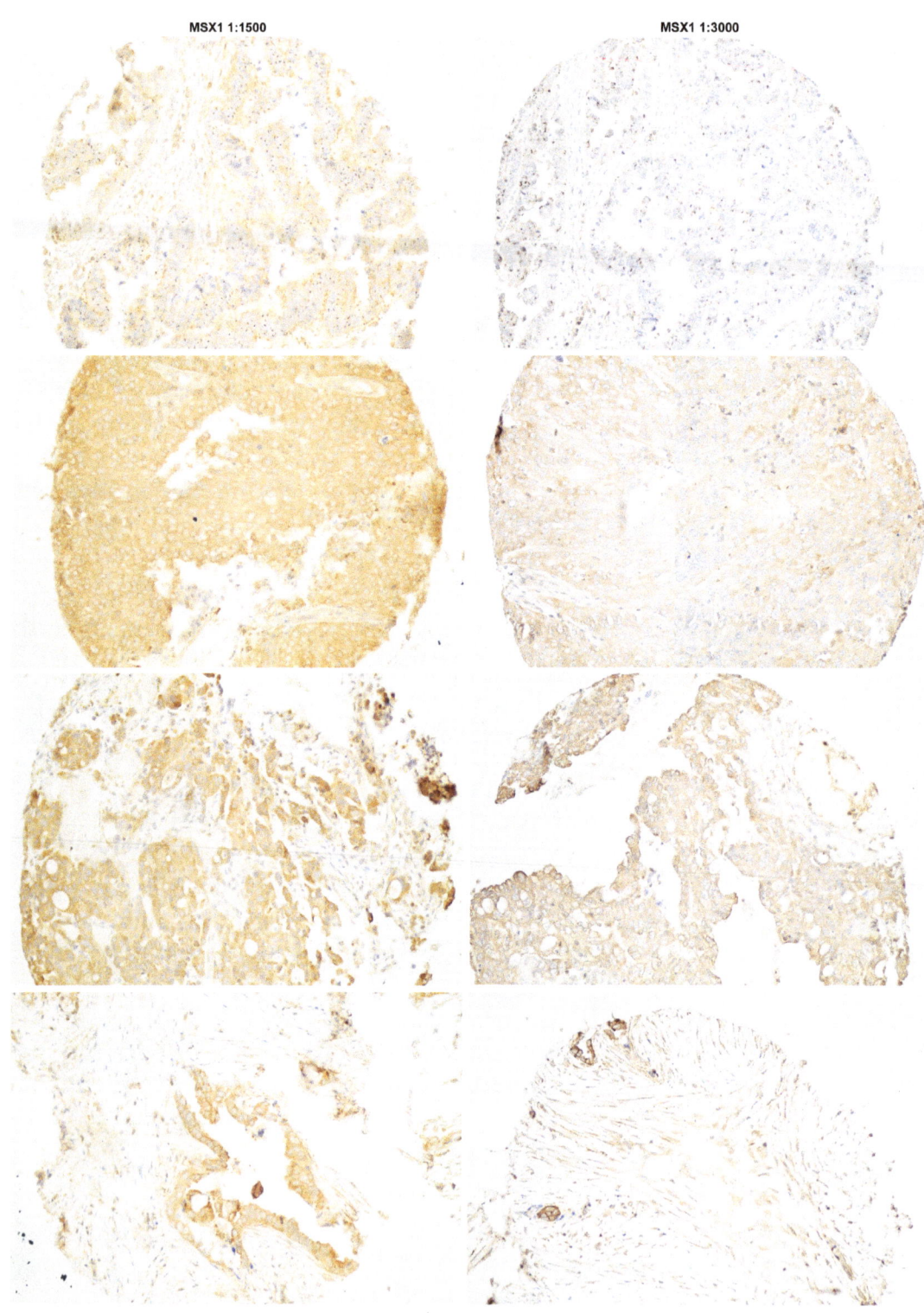

Two dilutions of primary antibody giving different results that reveal potential differences in protein expression between samples.

Co-localisation methods for immunohistochemistry.

In order to identify which specific cells show immunoreactivity for a protein target, it is possible to use thin serial sections, each stained for a different antigen, and then compare the staining in each section. A proportion of the cells should be present in both sections, providing that the sections are thin enough.

I used this method to co-localise neuroendocrine peptide with pituitary hormones, as part of my PhD. Pairs of 2μm serial sections of pituitary glands were mounted on poly-L-lysine coated slides. Pairs were either mounted on the same slide to enable the observer to move from one section to the other easily, or on separate slides to enable the observer to use two adjacent microscopes to view both sections "simultaneously". One section from each pair was stained for a pituitary hormone and the other for the novel peptide whose localisation was unknown, with antisera at optimum dilution and using a high resolution immunocytochemical method.

An alternative to the serial section method was to collect "mirror-image" pairs of paraffin sections. The first section of each pair was inverted before floating onto the water bath and picked up on the slide upside-down. The second section was collected in the conventional manner. This method permitted the identification of the same cell in both sections because cells which had been bisected by the microtome knife were present in both halves of the pair, and the sections did not have to be as thin as 2μm.

Some examples are shown on the next page. Both are from sections of rat pituitary stained for a peptide and a pituitary hormone which is expressed in the same cells. In order to obtain these results, sections were stained for the peptide alongside each of the pituitary hormones.
In each case, the peptide staining is on the left and the relevant co-localised hormone is on the right.

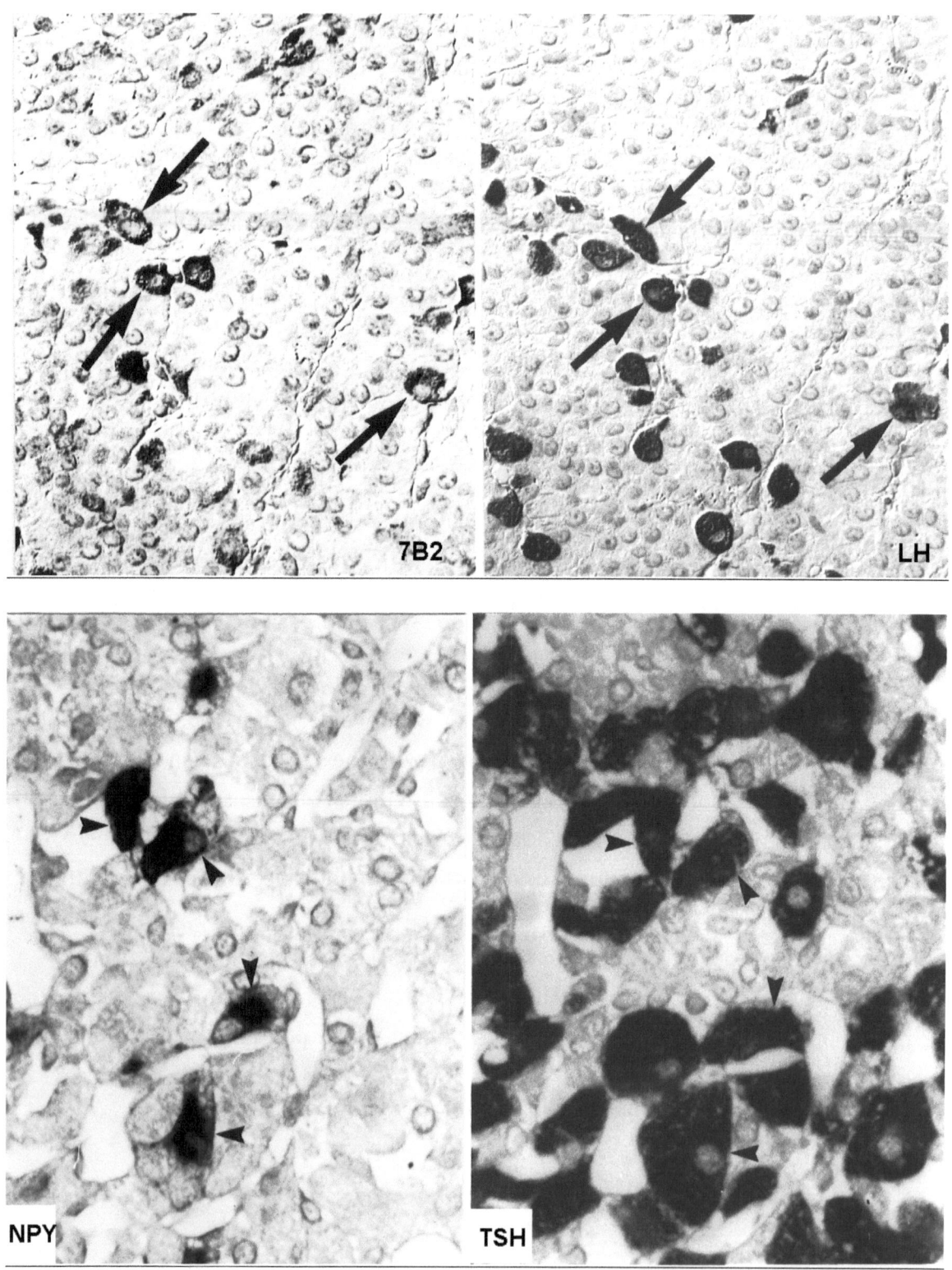

Co-localisation of 7B2 with LH and NPY with TSH in serial sections of rat pituitary. Arrows show some cells expressing both antigens.

In situ hybridisation with digoxigenin-labelled probes

Simple outline of procedure (see detailed protocol several pages further on):

Probe.

Probe design (antisense for detection of mRNA and sense controls if required). Probe length can be from 100 bases up to over 1 kb, if necessary. Optimum length about 200-500 bases.

RT-PCR for amplification of appropriate sequence.

For riboprobes: cloning of PCR product into vector containing RNA polymerase promoters (SP6, T3 or T7).

Linearisation of vector to generate linear template by digestion with restriction enzymes.

Probe labelling – in vitro transcription with cold NTPs and digoxigenin-UTP, using a suitable kit such as Ambion Megascript.

Estimation of probe yield using agarose gel electrophoresis.

You now have a digoxigenin-labelled cRNA probe ready for use.

Hybridisation.

Tissue sections or cells – paraffin sections, frozen sections, or cultured cells grown on plastic chamber slides. Fixation preferably in formalin or paraformaldehyde. Archive material can be used, or freshly prepared tissue blocks.

Designing experiment - selection of appropriate negative and positive controls, labelling slides.

Deparaffinising wax sections.

Permeabilising wax sections – digestion with proteinase K at 37°C.

Inactivation of proteinase K – treatment with 0.2% glycine.

Post-fixation – 4% paraformaldehyde.

Rinsing and air-drying sections.

Application of labelled probe diluted in hybridisation buffer.

Coverslips (Parafilm) placed over sections, slides sealed in moist slide boxes.

Incubation overnight at about 55°C.

Post-hybridisation washes – SSC, RNase, requires a shaking incubator at 55°C or 37°C.

Detection of digoxigenin-labelled hybrids using anti-digoxigenin labelled with reporter (usually alkaline phosphatase).

Development of colour reaction using chromogenic substrate (usually NBT/BCIP).

Washing slides, counterstaining if desired, mounting.

You now have a microscopic preparation with your target mRNA labelled visually.

Time needed:

Probe labelling (from a linearised template) is usually done well in advance, takes a few hours and probes are usually precipitated overnight, followed by running an agarose gel the next day. Labelled probes can be stored in -20°C freezer for months or years.

Tissue pretreatment and probe application will take a couple of hours, probe hybridisation overnight, followed by a whole day for washes and immunohistochemical detection. Usually slides are incubated in developing solution overnight, so any result will probably appear on day 3 after probe application.

If necessary the whole procedure can be carried out in one week, or if probe is already labelled, in 3 days.

Making probes for in situ hybridisation.

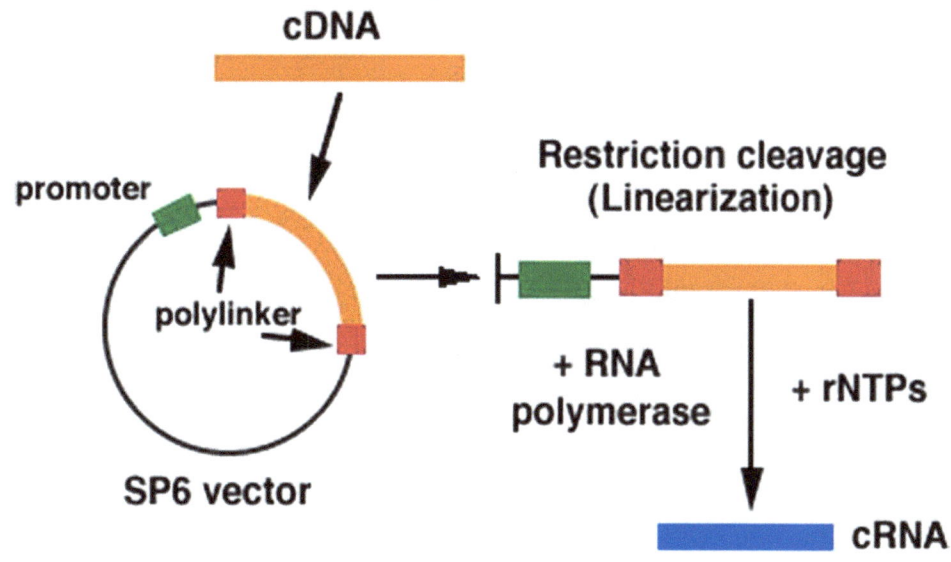

Hybridisation to messenger RNA.

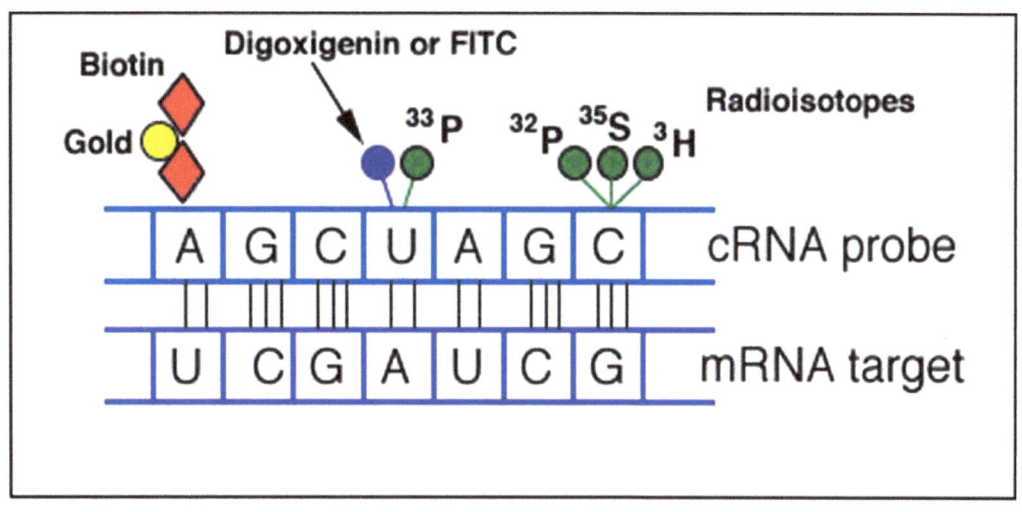

Why use digoxigenin?

The short answer is because it is the most sensitive non-radioactive label for RNA probes and in vitro transcription produces high specific-activity probes that should give a good signal-noise ratio when used to detect abundant RNAs. There are a variety of detection strategies available for digoxigenin-labelled probes, including fluorescent reporters. Digoxigenin is not expressed in animal cells and therefore endogenous digoxigenin will not be present to give non-specific labelling.

Most workers use digoxigenin although others use biotin as a label. Biotin can be found in some animal tissues and detection of biotin-labelled probes is usually less sensitive than digoxigenin, although there are various sensitive detection methods available.

Radioactive labels for ISH include ^{35}S, ^{32}P and ^{3}H and these methods can be very sensitive (1 silver grain over 1 cell can be a result). Some workers prefer to use radioactive labels because they can give higher contrast and greater sensitivity, but the expense, time sensitivity (half life) of radioactive isotopes, additional radioactive hazards, slowness of detection by autoradiography (sometimes weeks) and low cellular resolution must be weighed against the advantages of radiolabelling. Emulsion for coating slides is now becoming difficult to obtain commercially.

Digoxigenin-labelled probes can be stored at –20°C for months or years after labelling, unlike radioactively labelled probes which need to be labelled fresh each time they are used.

Equipment, reagents, solutions required.

Basic lab equipment

This is largely the same as for other histological or immunostaining methods.

For in situ hybridisation, it is necessary to have a set of equipment that is RNase free and is kept separate from the dishes used for non-RNase-free work.

Staining dishes (plastic or glass): one set RNase free, another set for RNase treatment and post-hybridisation washes.
Staining racks to hold slides.
Plastic trays where slides can be laid flat for probe application or staining methods.
Storage boxes for storing unstained and stained sections.
Plastic boxes for incubating slides for hybridisation, which can be sealed to keep a moist environment.
Sterile pipette tips and eppendorf tubes for diluting probes.
Bottles for preparing buffers (sometimes in litre quantities).

Incubator that can be set to variable temperatures with shaking platform – for washes and digestion steps. Maximum temperature will be 55°C.

Solutions and reagents (some to be made up before use).

For deparaffinising wax sections: Histoclear or xylene, 100% ethanol, 70% ethanol.

Chemicals etc:
Proteinase K stock solution (10 mg/ml, Sigma P2308, 500mg)
RNase A stock solution (100 mg/ml, Sigma R5503, 1g)

For hybridisation buffer (see recipe later on):
Dextran sulphate (Sigma D8906, 50g),
Deionised formamide (Sigma F9037, 100 ml),
Denhardt's solution (Sigma D2532, 5x5ml),
Herring sperm DNA (Sigma D-7290, 5x1ml),

The enzymes and hybridisation buffer can be stored as aliquots in freezer.

Digoxigenin-11-UTP (Roche, cat. 11209256910, 250nmol, 1mM, 25µl)
Anti-digoxigenin-AP Fab fragments (Roche, cat. 11093274910)
NBT (Roche, cat. 11383213001, 300mg in 3ml)
BCIP (Roche, cat. 11383221001, 150mg in 3ml)

Glycerol gelatin mountant (Sigma GG1, 10x15ml).

Sterile distilled water (autoclaved).
PBS made with autoclaved water (use 10x PBS from lab stocks at IRDB)
20xSSC (available from lab stocks at IRDB).

4% paraformaldehyde in PBS (made fresh on day of use).
0.2% glycine in PBS (made fresh on day of use).
20% acetic acid in methanol (stored at 4°C).

Stock solutions and buffers for alkaline phosphatase detection (non-sterile, stored at room temperature);
5M NaCl
1M MgCl$_2$
1M Tris-HCl pH7.4
1M Tris-HCl pH9.5

****

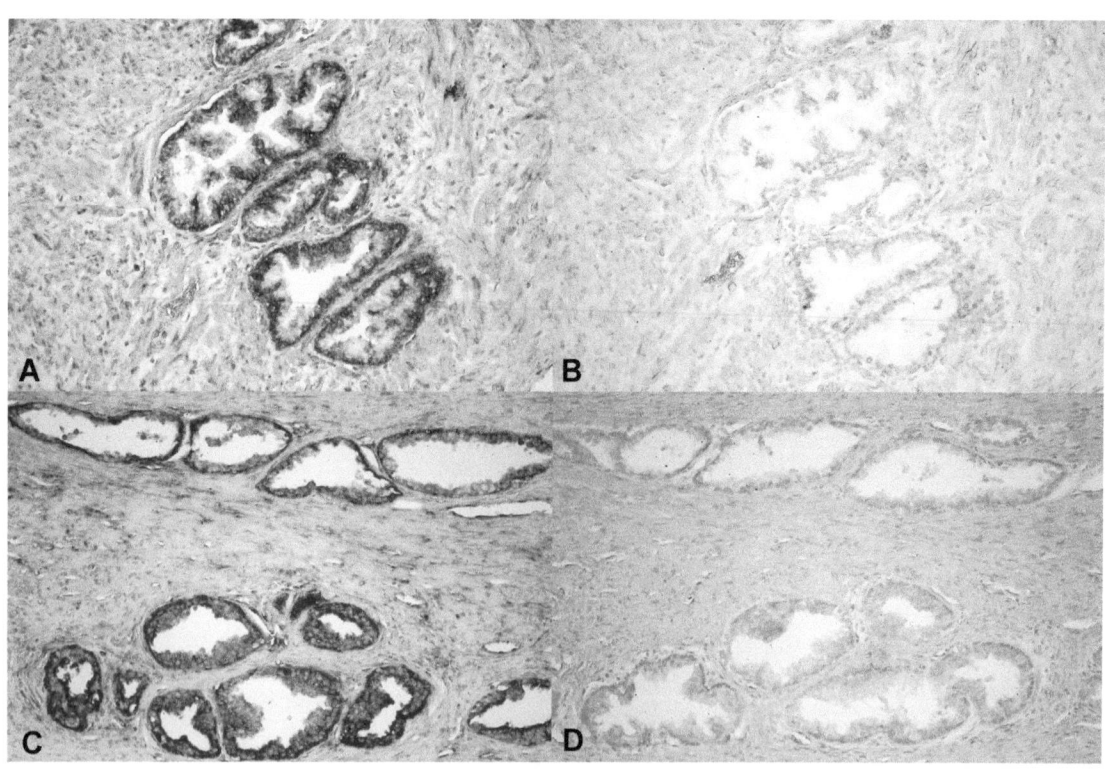

In situ hybridisation with RNA probes labelled with digoxigenin:
RIP140 expression in human prostate, (A) antisense probe, (B) sense probe. (C) untreated section and (D) RNAse-pretreated section hybridized with the antisense probe.

Controls for In situ hybridisation.

Appropriate controls must be included in every experiment (see figure on the previous page for examples). More details will be provided in the detailed protocol further on.

Positive controls:

As a general rule it is essential to include something that will definitely work in each experiment. It is a good idea to include a positive control probe such as beta-actin (appropriate species) which will provide some labelling in almost every tissue, in different types of cells according to its expression. The beta-actin labelling will indicate that there is sufficient good-quality mRNA present in the tissue to detect using in situ methods. Other positive controls may be more appropriate for certain tissues, e.g. prolactin for pituitary.

If testing a probe for the first time, use a positive control tissue that definitely contains the target. If there is no positive control for the new probe, include a method control (known positive tissue and another probe that usually gives good results.) This will provide an indication that the experiment has worked even if the test probe doesn't detect anything in the test tissue.

Negative controls:

These include
A) sections receiving no probe (hybridisation buffer only) to allow assessment of endogenous activity of e.g. alkaline phosphatase activity or autofluorescence,
B) RNase pre-treated sections (all mRNA removed before hybridisation),
C) sense probes (wrong orientation therefore should not bind to target)
D) inappropriate probes, for wrong species or for target sequences that are not expressed in the tissue. Negative labelling will provide evidence that the positive labelling obtained with a test probe is specific to that probe rather than binding of cRNA to tissue elements.

The use of sense probes can cause problems and sometimes a sense probe will inexplicably give high background labelling or even similar labelling to the antisense probe. Some workers avoid the use of sense probes completely and their work has been published without problems. The use of positive controls, such as other probes labelling a different target, and negative controls such as inappropriate probes or RNase pretreatment, are usually acceptable to prove that the antisense probe results are genuine.

For detailed background on the use of riboprobes for ISH see the following book:
Polak J.M, Wharton J, McGee J. O'D (eds). *In situ* Hybridisation: Principles and Practice, 2nd Edition, Oxford University Press, 1999. Chapter 4: Steel J.H, Gordon L, Polak J.M. Principles and applications of complementary RNA probes.

Detailed In Situ Hybridisation Protocol

This protocol is designed for use with Ambion Megascript kits for transcription of cDNA templates in suitable vectors containing promoters for RNA polymerases such as T7, T3 and SP6, to yield cRNA antisense riboprobes (or sense riboprobes, depending on the orientation of the template sequence in the vector). Other kits are available, but in my hands the Ambion kit produces a good yield of labelled cRNA probe, because it contains NTPs at higher concentrations than are available elsewhere. Alternatively, transcription can be carried out using all the basic reagents and making them up to appropriate concentrations.

Probe template preparation.

PCR product is cloned into a suitable vector containing T7, T3, or SP6 RNA polymerase promoter sites, eg. pCRII-TOPO, Bluescript. Preferably choose a vector with 2 sites so that the insert can be transcribed in either direction.
Miniprep of DNA should be cleaned up to remove RNase if possible.
Spectrophotometer used to measure DNA concentration.
Sequencing is useful to determine orientation of sequence in the vector.

The vector needs to be linearised to use as a template for in vitro RNA transcription, in order to make a probe.
Restriction enzymes are used to cut the vector distal to the insert, allowing the SP6 or T7 promoter to be used to transcribe an antisense or sense probe.

For example, in pCRII-TOPO, if the sequence is inserted in forward orientation:
Digestion with BamHI or SpeI will give a template for generation of an antisense probe from the T7 promoter.
Digestion using EcoRV or NotI will give a template for generation of a sense probe from the SP6 promoter.

Digest typical recipe, total volume 200 µl:
x10 buffer	20 µl
BSA	2µl (if required)
Vector DNA	5 µl (depends upon concentration, no more than 10 µg)
Enzyme	5 µl
Water	to 200 µl

Digest for 1-2 hours at 37°C.
Extract linearised template using phenol/chloroform/isoamyl, then chloroform alone.
Precipitate template using 30 µl of 5M ammonium acetate and 500 µl of absolute ethanol. Store at -20°C for at least 2 hours.
Centrifuge maximum speed for 5 min.
Resuspend in dH_2O, RNase free, in a volume appropriate to initial amount of DNA.
Template should be at 1 µg/µl if possible.

Digoxigenin riboprobe labelling.

1. Set up transcription reaction using an Ambion Megascript kit, appropriate to the RNA polymerase for the promoter present in each probe template; i.e. use SP6, T7, or T3 kit. Reagents from the kit are marked with an asterisk*, others must be obtained from other sources.

Typical transcription reaction made up as follows:

*10x transcription buffer	2 µl
*ATP, *CTP, *GTP (50 mM for SP6, 75 mM for T7 or T3)	2 µl each
*UTP (50 mM for SP6, 75 mM for T7 or T3)	1.3 µl
Digoxigenin-11-UTP (Roche 1209256, 250 nmol)	3.4 µl (for SP6) or 5 µl (for T7/T3)
DNA template (1 µg/µl)	1 µl
*Enzyme mix	2 µl
*sterile dH$_2$O	to total volume of 20 µl.

The ratio of final dig-11-UTP to unlabelled UTP concentration is 1:3 as recommended in the Ambion manual.

2. Incubate transcription at 37°C in water bath or incubator for 2-6 hours (the longer the better).

3. Remove a 1 µl sample of unpurified transcription reaction from each tube and add to a tube containing 10µl sterile water or TE buffer (10 mM Tris-HCl pH 7.6, 1 mM EDTA). Store at -20°C until the gel electrophoresis step (steps 9&10). This will be used to check that the transcription has worked and estimate recovery of labelled probe.

4. Terminate transcription of remaining 19 µl by addition of 1 µl DNase (RNase-free) per tube (2 U/µl, from Ambion kit or from Roche). Incubate at 37°C for 15 mins.

5. To precipitate the labelled probe, add 25 µl of lithium chloride solution from kit (7.5 M LiCl, 50 mM EDTA), plus 30 µl RNase-free water, plus 150 µl ice-cold ethanol to each tube. Mix well.
Place in -20°C freezer for at least 2 hours or preferably overnight to precipitate.

6. Centrifuge at maximum speed for 15-20 mins. Small pellet should be visible.

7. Remove supernatant by inverting tube or pipetting off carefully. Remove drops of fluid from neck of tube using clean paper tissue. Let pellet dry in air for a few minutes, does not need to be completely dry.

8. Resuspend pellet in 20 µl of TE.

9. Take a 1 µl sample of this resuspended purified transcript and add to 10 µl of sterile water or TE. Add 1 µl of gel loading dye from kit to each sample before loading gel. Do the same for the samples taken out before DNase treatment (step 3).

10. Run samples on 1-2% agarose gel in 1xTAE buffer. The sample taken out before adding DNase should have a high molecular weight template band plus a correct sized transcript band, the post-DNase sample should have only a band of correct size for purified transcript. In the image below, lane (A) shows pre-DNase and lane (B) post-DNase.
The recovery of labelled probe should be such that the amounts of RNA in each lane is similar. If more than one band is seen for the probe, this may be due to secondary structure (on a non-denaturing gel) and is not a problem.
Estimate quantity of yield (based on template band in pre-DNase sample, 1 µg added to original reaction, 1/20 of this run on gel = 0.05 µg).

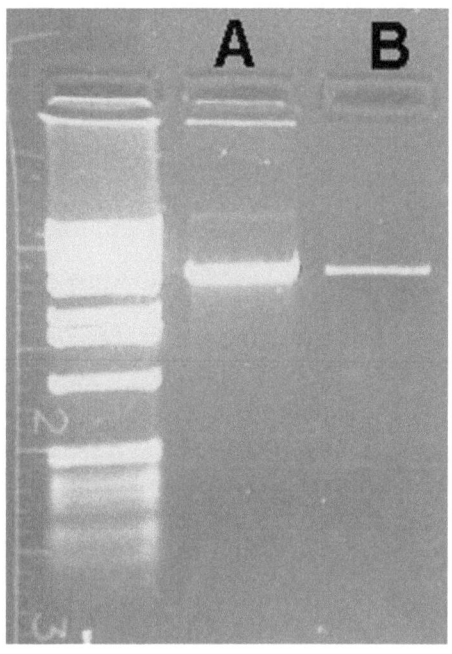

In situ hybridisation (ISH).

Paraffin or frozen sections should be mounted on slides that have been baked or are guaranteed RNase free, coated with adhesive such as poly-L-lysine or 3-aminopropyltriethoxysilane (Sigma cat no A3648). It is acceptable to buy ready-coated slides such as Polysine or Superfrost slides from Merck/BDH, no need to bake. When cutting paraffin sections, wear gloves and float sections out on autoclaved or fresh clean distilled water. Use a clean knife wherever possible. Dry slides at 37°C and keep them clean (store in a box when dry). Use paraffin sections fresh if possible or within a few days or weeks of cutting, as signal deteriorates in cut sections stored for long periods.

It can be convenient to mount 2 or more sections on each slide, spaced well apart, which can be hybridized with different concentrations of probe. This reduces the number of slides required and makes allowance for variation in experimental conditions which may affect labelling intensity.

Frozen sections can also be used, no proteinase K digestion required. Unfixed frozen sections should be dried for 1 hour and fixed in 10% neutral buffered formalin or 4% paraformaldehyde for 30 min, then go straight to step 6.
If using cultured cells, remove medium, fix cells in formalin for about 30 min, go straight to step 6.

Slides can be incubated in solutions contained in glass or plastic containers, which can be autoclaved or baked. Once used for RNase-free solutions these containers can be kept separate and re-used many times. Keep separate containers clearly marked for the RNase incubation step and do not allow RNase to contaminate any other equipment.

Make up all pre-hybridisation solutions using RNase-free (autoclaved) water. PBS can be made from 10x concentrate using autoclaved water, or use autoclaved PBS.
After the overnight probe hybridisation step, there is no need for RNase-free conditions because all target RNA will be hybridized, therefore double stranded, and immune to RNase digestion.

Before beginning the procedure, make up 4% paraformaldehyde, which needs to be made up fresh, dissolved in warm PBS (add a few drops of 1M NaOH if difficult to dissolve) and allow to cool on ice or in a fridge for a few hours.

Pre-hybridisation procedure:

1. Dewax paraffin sections through alcohols:
Histoclear (2 changes), 100% alcohol (2 changes), 70% alcohol, autoclaved water.

2. Rinse in phosphate-buffered saline (PBS).

3. Digest with proteinase K at 1-10 μg/ml in PBS at 37 °C, in order to permeabilise formalin-fixed tissue.
Immerse slides in a suitable volume of solution from about 5- 30 mins in a shaking incubator.
The appropriate proteinase K concentration and time of digestion will need to be determined for each type of tissue.
As a guide, 10 μg/ml for 10 min usually works well for most formalin-fixed mouse tissues. This may need to be adjusted, usually by increasing or decreasing the digestion time slightly. Over-fixed archive human tissue specimens may require more digestion, especially if originally fixed for several days.
Proteinase K stock solution: 10 mg/ml, add 100 μl per 100ml to obtain 10 μg/ml final concentration (store at –20°C).

4. Arrest proteinase K digestion by immersing in 0.2% glycine in PBS for 5 mins at room temp. Make up 0.2% glycine fresh (or from 2% stock, autoclaved and diluted 1:10).

5. Fix sections in freshly prepared cold 4% paraformaldehyde in PBS for 5 mins. Neutral buffered formalin can be used if PFA is not available.

6. Rinse in PBS, then in autoclaved water.

7. Air dry sections; drain slides and lay on rack in a square plastic tray horizontally to facilitate drying.

8. RNase-A treated negative controls can be done at this stage. This is a necessary control for checking specificity of probes. There is no need to do every time, but a thorough study should include this step once at least.

RNase-A treatment: following proteinase K digestion, glycine and PFA treatment, take slides out at the PBS stage for RNase-A digestion and keep separate from all the other slides from now onwards to prevent contamination of RNase-free test slides.

Rinse slides in 2xSSC, then digest with 100 μg/ml RNase-A in 2xSSC for 1 hour at 37°C with agitation. Rinse well in 2xSSC, then in PBS. Fix briefly in 4% paraformaldehyde to destroy enzyme. To remove all the RNA in sections may take extensive digestion, so concentration or digestion time may need to be increased if any signal remains when the probe is applied. Eventually, it will be possible to remove all the RNA.
Rinse slides in autoclaved water and air dry. Apply probe as normal (see below).

Hybridisation procedure

1. Make up diluted antisense and sense probes and positive control probes using hybridisation buffer (see "Buffers" below), warmed to 50-55°C before use. Assume you will need 10-15 µl probe per section in calculating how much to make up, and include extra quantity to allow for loss during pipetting.
Incubate diluted probes in water bath at 50-55°C for a few minutes before use. Usually, optimum concentration of a new probe will be unknown, so try serial dilutions of the stock solution, 1 µl of which was run on the gel. Most probes seem to work well at dilutions from 1:100 (i.e. add 1 µl of probe stock to 100 µl of hybridisation buffer to start with), and some can be diluted as much as 1:1000. If bands on gel were very weak, start at 1:50.

2. Apply about 10-15 µl of diluted probe to each section, depending on the section size. Place the drop of probe just beside the section.
Include sections which receive buffer only as negative controls, or sense probe controls. Remember that a small volume of probe can be spread thinly to cover a large section, so don't add too much.

3. Quickly cover sections with squares of Parafilm, cut to the right size to cover the section. Before adding the probe to the sections, cut enough pieces of Parafilm so that these are ready for use.
Ensure the whole section is covered with solution and there are no large air bubbles – press gently with forceps to remove bubbles. If the Parafilm is not completely flat, it will flatten in the oven at 55°C. Some workers prefer to use baked glass coverslips, which work equally well but are of fixed size and more difficult to remove. Parafilm has the advantage of being cut to fit the section and will float off when slides are immersed in buffer. Providing it is handled only with gloved hands and a clean supply is available, Parafilm does not appear to pose a risk of RNase contamination.

4. Place completed slides in a moist chamber, humidified with autoclaved water, and seal with tape. Slides must be supported horizontally to prevent coverslips sliding off, so use flat supports in a sandwich box or slide storage boxes stood on end. Slide boxes where slides are usually stored vertically can be stood on their ends, so that the slides are horizontal.
Place moist chambers in hybridisation oven or incubator (preferably one with a fan and good temperature control) at 55°C overnight. Temperature of hybridisation is important and can be adjusted. For small probes, lower temperatures may be used, but 55°C has worked well for many probes of 200 bases to 1.6 kb. For oligonucleotide probes, use 37°C.

Post-hybridisation washes and detection of bound probe.

1. Next day, remove Parafilm coverslips by immersing slides in slide racks in 2x SSC buffer at room temperature. Make up 2xSSC from stock 20x SSC using distilled water (not autoclaved) - see "Buffers" below.

Care to avoid inhaling toxic vapour from the formamide in the hybridisation buffer - it may be best to use a fume hood for removing slides from moist chamber and placing in buffer. Otherwise, make sure that the slides are cooled to room temperature before removing from slide boxes.

2. Wash slides at room temp in 2x SSC (2 changes). Slides can stay in 2x SSC for long periods if necessary.

3. To remove unhybridised single-stranded digoxigenin-labelled RNA probes, digest with RNase-A at 100 µg/ml in 2x SSC at 37°C for 30 mins. The time can be increased if there are severe problems with background.
RNase-A stock solution: 100 mg/ml made up in distilled water (store at –20°C). Add 100 µl per 100 ml of buffer to obtain 100 µg/ml.

4. Rinse slides briefly in 2x SSC at room temp.

5. Wash slides at 50-55°C in 2x SSC (2 changes) and 1x SSC, about 30 mins per wash. This can be reduced to 2 changes of 10 min each if pressed for time.
The stringency can be increased if desired by diluting to 0.5x SSC or even 0.1x SSC if required. These washes should be carried out with agitation if possible. Timing is not critical.

6. Rinse slides in AP1 buffer, pH 7.6 (see "Buffers" below for recipe) at room temperature. TBS can also be used at this stage. No time limit, but 5 mins is enough.

7. Block endogenous alkaline phosphatase by immersing slides in 20% acetic acid in methanol at 4°C for 30 seconds to 1 minute (this blocks intestinal alkaline phosphatase, but may not be necessary for other tissues. If in doubt, do it anyway). Acetic acid/methanol can be stored in fridge and re-used several times. Do this step in a fume hood.

8. Rinse slides in AP1 buffer to remove traces of methanol and acetic acid.

9. Wipe carefully around sections, leaving section wet but remainder of slide dry. Lay flat in moist chamber.

10. Block non-specific binding by incubating with AP1 buffer containing 3% bovine serum albumin (AP1/BSA), 30 mins at room temp.

11. Drain sections (do not rinse off AP1/BSA) and wipe away excess liquid.

Apply a few drops per section of anti-digoxigenin-AP conjugate (Roche), diluted to 1:500 in AP1/BSA. Leave to incubate with antibody for 1 hour at room temp.

12. Rinse slides in AP1 buffer, pH 7.4, 2 or 3 changes.

13. Rinse in AP2 buffer, pH 9.2 (see "Buffers" for recipe).

14. Drain slides, wipe around sections and apply a small volume of developing solution to slides laying flat in moist chamber.

Make up developing solution from components as follows -
Stock solutions (all stored at -20°C):
A) 100 mg/ml nitro blue tetrazolium (NBT) in dimethylformamide (DMF) (from Roche, cat no 1383213).
B) 50 mg/ml 5-bromo-4-chloro-3-indolyl phosphate (BCIP), in DMF (from Roche, cat no 1383221).
C) AP2 buffer plus 1 mM levamisole (25 mg levamisole per 100 ml). Can be made fresh or made in advance and stored in freezer.

Working solution:
add 35 µl of the NBT solution (A) and 35 µl of the BCIP solution (B) to 10 ml of the AP2/levamisole solution (C).

Alternatively, dissolve one Sigma NBT/BCIP tablet in 10 ml distilled water, or use Vector alkaline phosphatase substrate kits (various colours including NBT/BCIP, Vector Blue and Fast Red). However, the best results so far have been obtained using the individual NBT/BCIP reagents as above.

15. Place slides in darkness (wrap in foil or place in a cupboard) at room temp and leave to incubate in developing solution for as long as required. Check microscopically initially after 1-2 hours. If no reaction has occurred, leave slides for much longer or overnight. In practice, if stain does not appear after 1-2 hours, overnight incubation is usually necessary. If the stain has been very intense in a previous experiment, the incubation time may need to be adjusted to avoid too much non-specific background labelling.

16. When sufficient colour has developed, rinse slides in AP2, then in distilled water and mount in glycerol gelatine (Sigma or Dako), or another aqueous mountant.
DO NOT dehydrate or expose to ethanol, xylene or Histoclear, because this will dissolve the NBT/BCIP reaction product.
Slides can be counterstained using Mayer's haematoxylin or Nuclear fast red, etc. if desired, before mounting.

17. If using Vector NBT/BCIP kit, the slides can be dehydrated, cleared and mounted in DPX. Some change in colour of reaction product may be seen when the slides are dehydrated.

18. Do not leave stained sections in bright sunlight. NBT/BCIP reaction product is still light-sensitive and sections may turn black if exposed to too much sunlight. Store slides in closed boxes. The completed slides can be kept for many years if stored correctly.

Buffers.

Hybridisation buffer recipe:

For 10 ml of buffer:
Weigh out 1 g of Dextran sulphate (Sigma D8906).
Dissolve at 50°C in
5 ml deionised formamide (Sigma F9037),
2.5 ml 20xSSC (must be RNase free, so autoclaved),

When fully dissolved, add
200 μl 50 x Denhardt's solution (Sigma D2532),
100 μl 10 mg/ml denatured sheared herring sperm DNA (Sigma D7290)
(DNA may need to be denatured in boiling water before adding to buffer).

Make up to 10ml final volume using autoclaved distilled water.
If using the same day, can be kept at room temp until required or store 1 ml aliquots in -20°C freezer. Heat to 55°C before use.

Final concentrations of reagents in this buffer are as follows:
50% formamide, 10% dextran sulphate, 1 x Denhardt's solution, 5 x SSC, 0.1 mg/ml herring sperm DNA.

20x SSC (3M NaCl, 0.3M sodium citrate, pH 7.0):

For 1 litre:
175.3g NaCl
88.23g Tri-sodium citrate
Dissolve in 800 ml distilled water, pH to 7.0, make up to 1 litre.
If using for hybridisation buffer, autoclave before use.
For post-hybridisation washes, no need to autoclave.

TAE buffer recipe for running agarose gel:

1x: 0.04M Tris-acetate, 0.001M EDTA;
Made up as 50x concentrate:
242g Tris base, 100 ml 0.5M EDTA, 57.1 ml glacial acetic acid, pH 8.0, made up to 1 litre.
Before use, dilute 20 ml to 1 litre with distilled water. It is not usually necessary to use RNase free water, although this can be used.

Buffers for alkaline phosphatase detection:

TBS

10x concentrate:
81 g sodium chloride, 6 g Tris base,
made up to 1 litre with distilled water and pH 7.6 with HCl.

Dilute x10 with water before use.

AP1 buffer

Make up from stock solutions as follows:

1 M Tris-HCl pH 7.4	100 ml
5 M NaCl	20 ml
1 M MgCl$_2$	2 ml

Make up to 1 litre with distilled water.

AP2 buffer

Make up from stock solutions as follows:

1 M Tris-HCl pH 9.2	100 ml
5 M NaCl	20 ml
1 M MgCl$_2$	50 ml

Make up to 1 litre with distilled water.

Examples of ISH results:

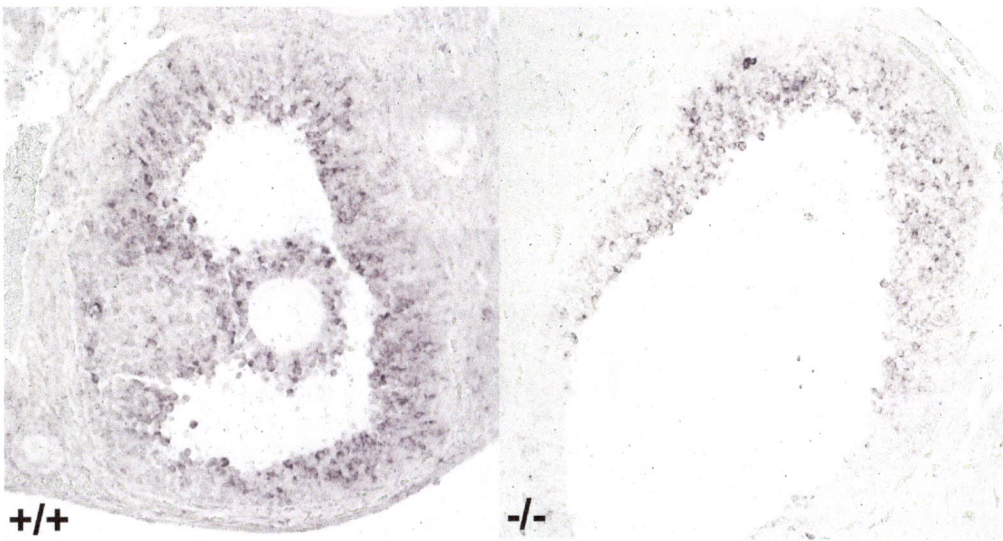

In situ hybridisation for amphiregulin mRNA, with antisense RNA probe labelled with digoxigenin. This shows the difference in localization and expression of amphiregulin between a RIP140 WT (+/+) and RIP140 KO (-/-) mouse, in cumulus cells in the ovary. Reference: Nautiyal et al, Endocrinology 151:2923-2932, 2010.

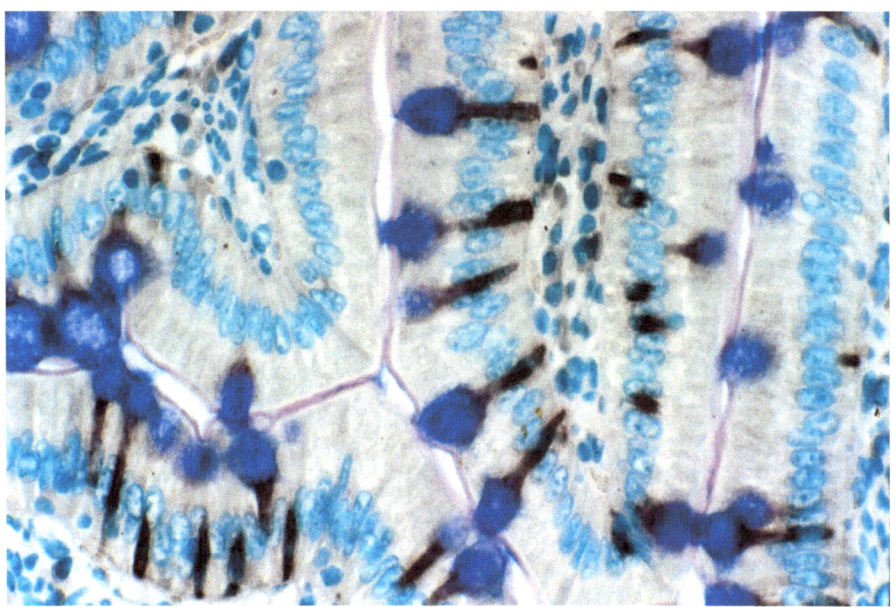

In situ hybridisation for ITF mRNA in mouse goblet cells (brown staining), section counterstained with alcian blue for the mucins in the goblet cells and pale blue counterstain for the cell nuclei.

Troubleshooting for In Situ Hybridisation:

Problem	Reason	Solution
No staining (section negative)	Probe concentration too low.	Try higher probe concentration.
No staining (section negative)	Proteinase K digestion insufficient.	Increase digestion time. Use positive control tissue and probes to check proteinase K conditions.
Whole section stained very intensely.	Probe concentration too high.	Dilute probe at least x2 next time.
Tissue structure is poor, cells disintegrating, vacuolation or poor nuclear detail.	Over-digestion with proteinase K.	Reduce digestion time.
Intense purple staining in acellular layers in some tissues.	Collagen tends to stain non-specifically in some tissues with NBT/BCIP.	Difficult to solve. Diluting probe or reducing detection time may help.
Negative control sections digested with RNase are still stained.	Inadequate RNase digestion or probe is not binding to RNA.	Try increasing RNase concentration or digestion time.
High background	Probe binding non-specifically.	Increase stringency of washes (dilute SSC or add 50% formamide to washes). Dilute probe further, or reduce detection time.
Sense probe gives positive labelling	Non-specific binding of sense probe.	Try diluting probe, increasing stringency of washes. If no improvement, rely upon other controls instead.
Buffer-only controls have staining.	Endogenous alkaline phosphatase present or tissue components that react with the colour substrate (eg. Collagen).	Try increasing blockade of alkaline phosphatase. If still present, will need to make allowance for this staining when examining positive sections.

Combined in situ Hybridisation (ISH) and Immunohistochemistry (IHC).

IHC and ISH can be combined on the same tissue section, allowing cells expressing mRNA to be compared with cells immunoreactive for the protein transcribed from that mRNA, or another protein. The two methods can be applied to separate sections, whether from different blocks fixed optimally for the two methods or serial sections of the same block fixed in a fixative which provides good mRNA and peptide preservation. However, there are problems associated with the use of separate sections, the main one being cell identification. Therefore, the use of both techniques on the same section is preferable. This can be carried out either with ISH or IHC first. There are advantages and disadvantages to both protocols, as described below. In my hands, hybridising mRNA prior to immunostaining gave better results, since there was no loss of mRNA which may occur if immunostaining precedes hybridisation.

<u>IHC followed by ISH.</u>

If IHC is performed first there will probably be optimal detection of antigens, but there is a risk of the mRNA being destroyed as a result of RNase contamination in antisera and non-immune sera which are used, and by handling the slides. Inclusion of RNase inhibitors in the antisera is the best way to avoid target degradation.
RNase inhibitors which have been used successfully include
Heparin, 1000-5000 U/ml diluted antiserum or DEPC, 0.04%.

I chose to use heparin, which was added to normal goat serum for preincubation of sections, and to secondary antibodies and other reagents antisera. The reaction was developed in DAB to give a brown reaction product. The slides were not dehydrated, but after washing in distilled water they were processed for in situ hybridisation starting with proteinase K treatment. The ISH method was carried out as usual.

<u>ISH followed by IHC.</u>

The advantage of carrying out the ISH first is that there will be optimal retention of mRNA and maximum detection of nucleic acids. However, the antigens in the cytoplasm which remain to be detected by IHC may be damaged by the hybridisation method. Care must be taken not to overdigest the tissue with proteases, or to overheat sections, which might also damage or denature the antigens. Dextran sulphate can impair antigenicity, and some workers suggest excluding it from the hybridisation buffer. Post-hybridisation washing can be reduced in order to avoid loss of antigens, then the slides were rinsed in PBS, before starting the immunocytochemical method with blocking of endogenous peroxidases. No RNase inhibitors were necessary in the antiserum incubations, since all mRNA was either stable and hybridised or removed by the RNase rinse.

Controls.

As with each technique carried out separately, controls are necessary to ensure the specificity of the reactions. In this case, it was first determined that in situ hybridisation and immunocytochemical methods independently gave specific, reproducible results on separate sections. Negative controls for each method were also included.

The example below shows rat pituitary with ISH for prolactin followed by IHC for TSH, in control (normal) rat and thyroidectomised rat, where the TSH cells increase in number and prolactin cells decrease. The ISH is purple-black and the IHC is brown.

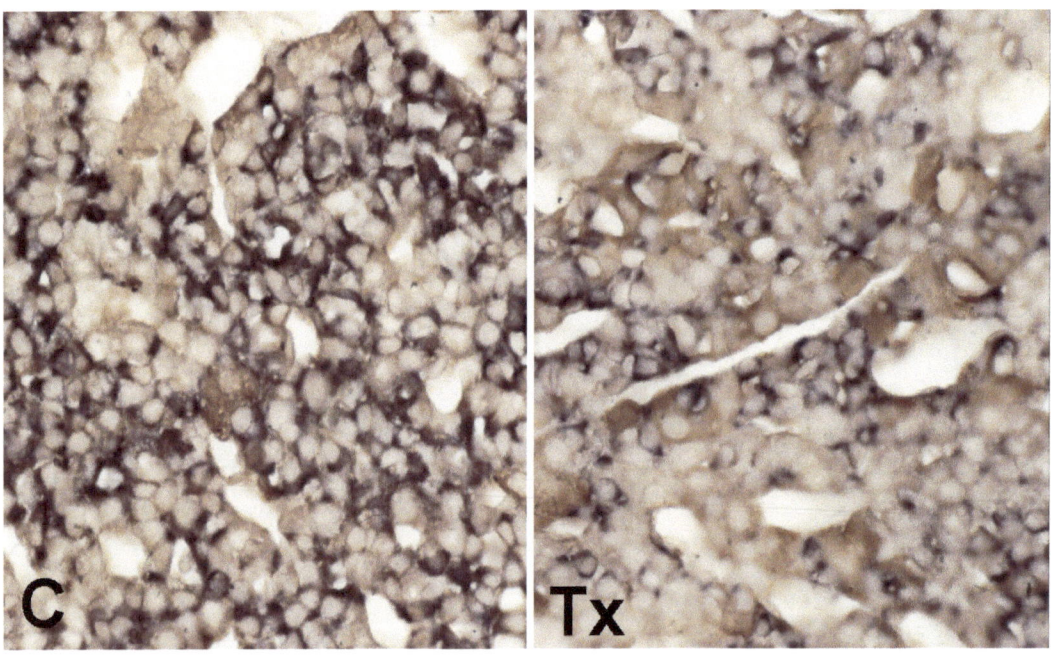

In Situ RT-PCR.

In the 1990s, a new technique was developed which attempted to revolutionise the detection of mRNA and DNA in tissues by microscopical methods. The aim of the technique is to amplify low copy number DNA or mRNA in cells or tissue sections, with incorporation of a label such as digoxigenin during the PCR step, so that the amplification product can be seen microscopically. From a mRNA target, reverse transcription (RT) is performed on the cells or sections, to produce cDNA. This is then used as a template for PCR incorporating digoxigenin-labelled dUTP. In the absence of RT, there should be no amplification product. DNase pretreatment is required to remove all nuclear DNA because if this is not done, the PCR reaction incorporates all the labelled nucleotide into the endogenous DNA rather than the newly-formed cDNA. I spent 4 years at Imperial Cancer Research Fund (ICRF) laboratories working on this method and attempting to get some meaningful results. I used galanin expression in the rat pituitary and small intestine as my target, because this peptide is expressed at relatively low levels and I thought that it had scope for amplification.

RT-PCR amplification using tissue sections mounted on glass slides was very problemmatic because the whole slide had to be heated to various temperatures, on a flat hotplate PCR machine, and the sections had to be sealed into a chamber holding the reagents, made from plastic coverslips and rubber cement. Many failed experiments using expensive reagents took place.

Eventually, I decided to use a different method where sections of a small paraffin block made from rat pituitary cell reaggregates were mounted onto pieces of round glass coverslips and immersed in 0.5 ml PCR tubes. The tubes were then inserted into the heating block of a PCR machine and the issues of uneven heating and loss of liquid from slides were avoided. This was a much more economical method and produced some results.

The results were published in a paper in 2001.

Steel J.H, Morgan D.E, Poulsom R. Advantages of *in situ* hybridisation over direct or indirect *in situ* RT-PCR for localisation of galanin mRNA expression in rat small intestine and pituitary. Histochemical Journal 33:201-211, 2001.

The abstract reads as follows:
"In situ hybridisation (ISH) and direct or indirect in situ reverse transcriptase-polymerase chain reaction (RT-PCR) were used to detect galanin mRNA in paraffin sections of rat intestine and pituitary. With conventional ISH, a subset of intestinal neuronal ganglion cells and anterior pituitary endocrine cells were labelled. Direct in situ RT-PCR also labelled some cells in pituitary but not in intestine. Negative controls were unlabelled, but sections with 3' primer alone for RT-PCR appeared positive. No signal was apparent using the indirect in situ RT-PCR method. Investigation of the specificity of solution phase RT-PCR using RNA extracts from pituitary or intestine revealed that additional PCR products were detected under some conditions. The sequences of these PCR products suggested that one was the result of mispriming and single primer PCR, which could also have occurred in situ. Alternative galanin

primers gave only the predicted RT-PCR product in solution phase yet still gave artefacts in tissue sections using direct in situ RT-PCR. ISH with probes transcribed from the correct PCR product gave identical labelling to the original galanin riboprobe. Conclusion: direct in situ RT-PCR is unreliable and requires validation, while indirect in situ RT-PCR may fail even though sufficient target exists for detection with conventional sensitive riboprobe ISH."

The figures below illustrate the results:

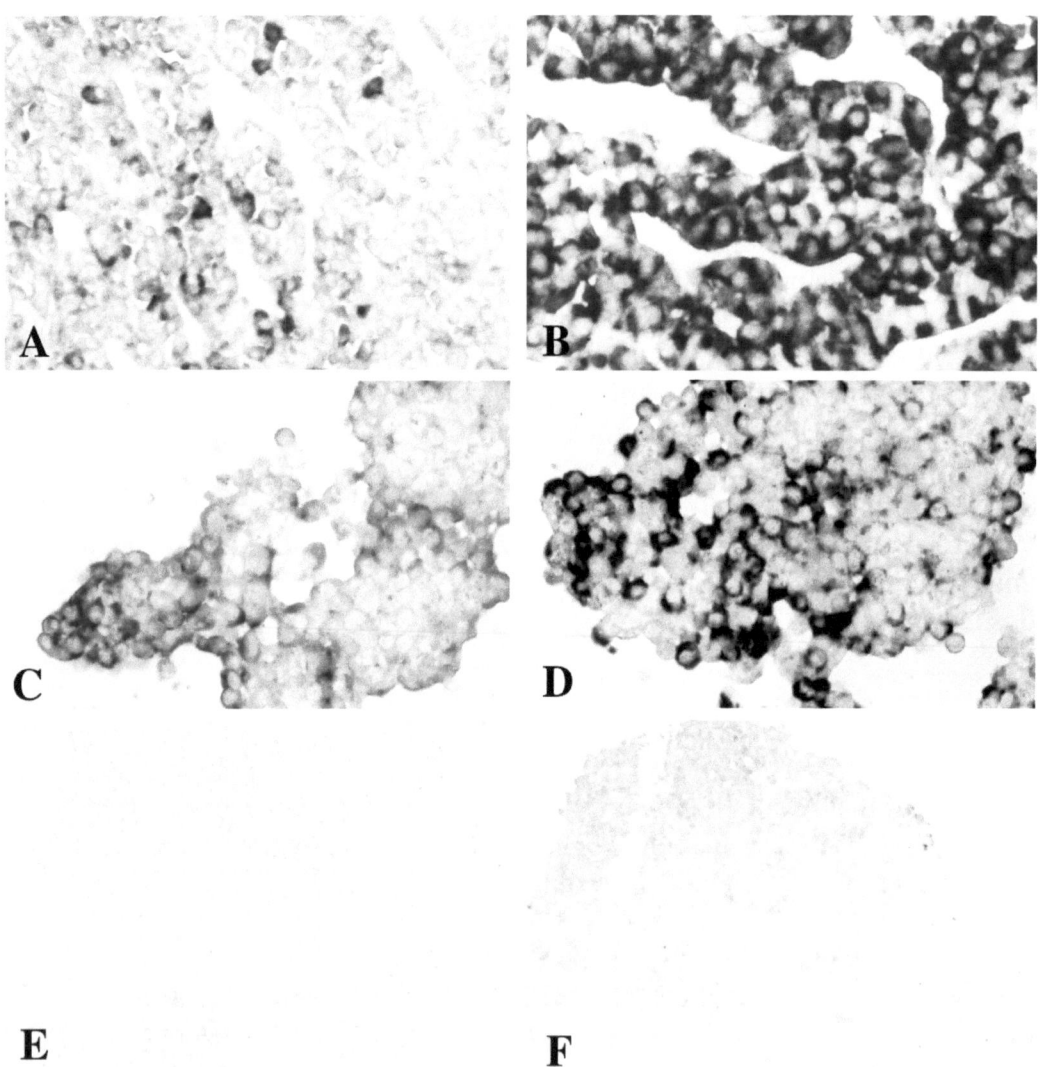

E   F

Figure 1: In situ hybridisation for galanin (A) and prolactin (B) mRNAs in rat pituitary (oestrogen treated female) and for galanin (C) and prolactin (D) mRNAs in pituitary cell reaggregates, detected using digoxigenin-labelled cRNA probes.
Negative control sections of pituitary (D) or reaggregates (E) that received buffer only. Galanin mRNA is expressed in scattered cells in the oestrogen-treated pituitary and equivocally in the reaggregates. In contrast, prolactin mRNA is abundantly expressed in both types of tissue.

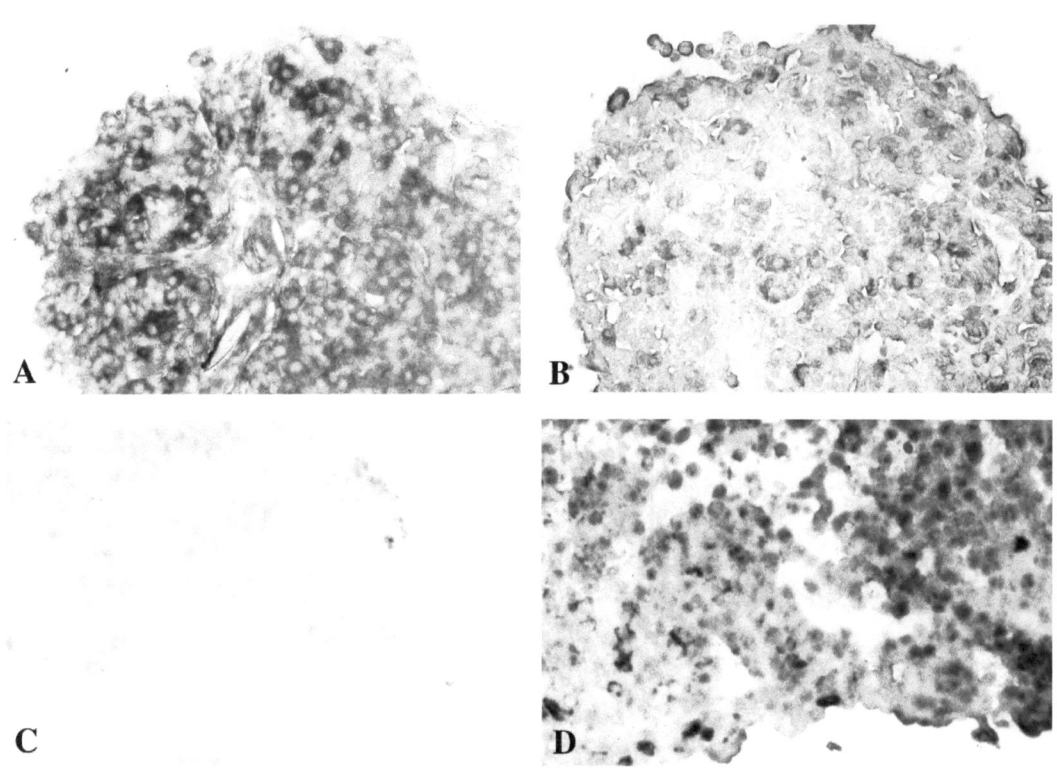

Figure 2: In situ RT-PCR with galanin primers on rat pituitary cell reaggregates. In DNase-pretreated sections,
(A) cytoplasmic labelling with negative nuclei in sections that had RT with 3' primer and PCR with both primers;
(B) similar labelling in sections after RT with 3' primer but which had PCR with no primers added;
(C) no labelling in sections after PCR with both primers but omitting the RT step. (D) In sections which had no DNase, primer-independent nuclear labelling only.

We concluded that it was possible to show a certain amount of amplification of the galanin target mRNA, which was already detectable in this tissue using ISH methods. The negative control sections with no RT had no reaction product, confirming that the labelling had come from a mRNA target converted to cDNA and then amplified by directly labelled PCR. The problem was that there were artefacts from inappropriate reaction products in some controls and that ISH alone worked more reliably.
In situ RT-PCR is potentially useful for detection of low copy number mRNAs which are difficult to detect using conventional ISH, but extensive validation is required to ensure that the results are not artefacts. In fact, for a target that cannot be detected using ISH, it would be impossible to validate the in situ RT-PCR unless by RT-PCR in solution phase.

Selected References which illustrate the use of histology, IHC and ISH:

Steel J.H, Van Noorden S, Ballesta J, Gibson S.J, Ghatei M.A, Burrin J, Leonhardt U, Domin J, Bloom SR, Polak JM. Localisation of 7B2, neuromedin B, and neuromedin U in specific cell types of rat, mouse and human pituitary, in rat hypothalamus and in 30 human pituitary and extra-pituitary tumors. Endocrinology 122: 270-282, 1988.

Steel J.H, Hamid Q, Van Noorden S, Jones P, Denny P, Burrin J, Legon S, Bloom S.R, Polak J.M. Combined use of *in situ* hybridisation and immunocytochemistry for the investigation of prolactin gene expression in immature, pubertal, pregnant, lactating and ovariectomised rats. Histochemistry 89: 75-80, 1988.

Steel J.H, Gon G, O'Halloran D.J, Jones P.M, Yanaihara N, Ishikawa H, Bloom S.R, Polak J.M. Galanin and vasoactive intestinal polypeptide are colocalised with classical pituitary hormones and show plasticity of expression. Histochemistry 93: 183-189, 1989.

Steel J.H, O'Halloran D.J, Jones P.M, Van Noorden S, Chin W.W, Bloom S.R, Polak J.M. Combined use of immunocytochemistry and *in situ* hybridization to study beta thyroid-stimulating hormone gene expression in pituitaries of hypothyroid rats. Molecular and Cellular Probes 4: 385-396, 1990.

Steel J.H, Martínez A, Springall D.R, Treston A.M, Cuttitta F, Polak J.M. Peptidylglycine a-amidating monooxygenase immunoreactivity and messenger RNA in human pituitary and increased expression in pituitary tumours. Cell and Tissue Research 276:197-207, 1994.

Steel J.H, Terenghi G, Hudson L.D, Chung J.M, Na S.H, Carlton S.M, Polak J.M. Increased nitric oxide synthase immunoreactivity in dorsal root ganglia in a rat neuropathic pain model. Neuroscience Letters 169:81-84, 1994.

Steel J.H, Jeffrey R.E, Longcroft J.M, Rogers L, Poulsom R. Comparison of isotopic and non-isotopic labelling for *in situ* hybridisation of various mRNA targets with cRNA probes. European Journal of Histochemistry 42:143-150, 1998

Steel J.H, Morgan D.E, Poulsom R. Advantages of *in situ* hybridisation over direct or indirect *in situ* RT-PCR for localisation of galanin mRNA expression in rat small intestine and pituitary. Histochemical Journal 33:201-211, 2001.

Leonardsson G, **Steel J.H**, Christian M, Pocock V, Milligan S, Bell J, So P.W, Medina-Gomez G, Vidal-Puig A, White R, Parker M.G. Nuclear receptor corepressor RIP140 regulates fat accumulation. PNAS 101:8437-8442, 2004.

Steel J.H, O'Donoghue K, Kennea N, Sullivan MHF, Edwards AD. Maternal origin of inflammatory leukocytes in preterm fetal membranes, shown by fluorescence in situ hybridization for X and Y chromosomes. Placenta, 26:672-677, 2005.

Steel J.H, Malatos S, Edwards A.D, Miles L, Duggan P, Kennea N, Reynolds P.R, Feldman RG, Sullivan MHF. Bacteria and inflammatory cells in fetal membranes do not always cause preterm labour. Pediatric Research, 57:404-411, 2005.

Tullet J.M.A, Pocock V, **Steel J.H**, White R, Milligan S, Parker M.G. Multiple signalling defects in the absence of RIP140 impair both cumulus expansion and follicle rupture. Endocrinology, 146:4127-4137, 2005.

Nichol D, Christian M, **Steel J.H**, White R, Parker M.G. RIP140 expression is stimulated by estrogen-related receptor a during adipogenesis. Journal of Biological Chemistry, 281:32140-32147, 2006.

Seth A, **Steel J.H**, Nichol D, Pocock V, Kumaran M.K, Fritah A, Mobberley M, Ryder T.A, Rowlerson A, Scott J, Poutanen M, White R, Parker M.G. The transcriptional corepressor RIP140 regulates oxidative metabolism in skeletal muscle. Cell Metabolism, 6:236-245, 2007.

Fritah A, **Steel J.H**, Nichol D, Parker N, Williams S, Price A, Strauss L, Ryder T.A, Mobberley M.A, Poutanen M, Parker M, White R. Elevated expression of the metabolic regulator RIP140 results in cardiac hypertrophy and impaired cardiac function. Cardiovascular Research, 86:443-451, 2010.

Nautiyal J, **Steel J.H**, Rosell M.M, Nikolopoulou E, Lee K, DeMayo F.J, White R, Richards J.S, Parker M.G. The Nuclear Receptor Cofactor RIP140 is a positive regulator of Amphiregulin Expression and Cumulus Cell-Oocyte Complex Expansion in the Mouse Ovary. Endocrinology, 151:2923-2932, 2010.

Salker M, Christian M, **Steel J.H**, Nautiyal J, Lavery S, Trew G, Webster Z, Al-Sabbagh M, Puchchakayala G, Foller M, Landles C, Sharkey AM, Quenby S, Aplin JD, Regan L, Lang F, Brosens J.J. Deregulation of the serum- and glucocorticoid-regulated kinase SGK1 in the endometrium causes reproductive failure. Nature Medicine, 17:1509-1513, 2011.

Nautiyal J, **Steel J.H**, Rosell Mane M, Oduwole O, Poliandri A, Alexi X, Wood N, Poutanen M, Zwart W, Stingl J, Parker M.G. The transcriptional co-factor RIP140 regulates mammary gland development by promoting the generation of key mitogenic signals. Development 140:1079-1089, 2013.

Magnani L, Patten D.K, Nguyen VTM, Hong S-P, **Steel J.H**, Patel N, Lombardo Y, Faronato M, Gomes A.R, Woodley L, Page K, Guttery D, Primrose L, Fernandez Garcia D, Shaw J, Viola P, Green A, Nolan C, Ellis I.O, Rakha E.A, Shousha S, Lam EW-F, Gyorffy B, Lupien M, Coombes RC. The pioneer factor PBX1 is a novel driver of metastatic progression in ERα-positive breast cancer. Oncotarget 6:21878-21891, 2015.

Hopkins TG, Mura M, Al-Ashtal A, Lahr RM, Abd-Latip N, Sweeney K, Lu H, Weir J, El-Bahrawy M, **Steel JH**, Ghaem-Maghami S, Aboagye EO, Berman AJ, Blagden SP. The RNA-binding protein LARP1 is a post-transcriptional regulator of survival and tumorigenesis in ovarian cancer. Nucleic Acids Research, Volume 44:1227-1246, 2016.

Blondrath K, **Steel JH**, Katsouri L, Ries M, Parker MG, Christian M, Sastre M. The nuclear cofactor receptor interacting protein-140 (RIP140) regulates the expression of genes involved in Aβ generation. Neurobiology of Aging 47:180-191, 2016.

Patel H, Abduljabbar R, Lai CF, Periyasamy M, Harrod A, Gemma C, **Steel JH**, Patel N, Busonero C, Jerjees D, Remenyi J, Smith S, Gomm JJ, Magnani L, Gyorffy B, Jones LJ, Fuller-Pace F, Shousha S, Buluwela L, Rakha EA, Ellis IO, Coombes RC, Ali S. Expression of CDK7, Cyclin H, and MAT1 is elevated in breast cancer and is prognostic in estrogen receptor-positive breast cancer. Clin Cancer Res 22:5929-5938, 2016.

www.ingramcontent.com/pod-product-compliance
Lightning Source LLC
Chambersburg PA
CBHW051056180526
45172CB00002B/656